I0029284

R. F.. Scharff

The History Of The European Fauna

R. F.. Scharff

The History Of The European Fauna

ISBN/EAN: 9783741103513

Manufactured in Europe, USA, Canada, Australia, Japa

Cover: Foto ©Klaus-Uwe Gerhardt /pixelio.de

Manufactured and distributed by brebook publishing software
(www.brebook.com)

R. F.. Scharff

The History Of The European Fauna

THE HISTORY OF THE EUROPEAN FAUNA

R. F. SCHARFF, B.Sc., Ph.D., F.Z.S.

*K. er of the Natural History Collections, Science and Art Museum,
Dublin; Member of the Royal Irish Academy; Corresponding
Member of the Senckenbergische Naturforschende
Gesellschaft.*

WITH ILLUSTRATIONS.

LONDON
WALTER SCOTT, LIMITED
PATERNOSTER SQUARE
1899

Acc. No.	2390
Class No.	B 21
Book No.	53

CONTENTS.

PREFACE.

Our knowledge of the present and past fauna of Europe is as yet insufficient to indicate with precision the original homes of its component elements, but I hope that the lines of research laid down here, and the method of treatment adopted, will aid zoologists and geologists in collecting materials for a more comprehensive study of the history of our animals. I trust also that a fresh impulse will be given by the publication of this book to the study of the Geographical Distribution of Species. Collectors of Beetles, Butterflies, Shells, and Fossils may derive some useful hints by its perusal and thus direct their studies, so as to add, by accuracy in observation, to our knowledge of the former geographical revolutions which have moulded our islands and continents. To geographers, a survey of some of the more important changes in the distribution of land and water in past times—based upon the composition of our fauna—will be interesting. The subject, however, is a complex one. I have ventured to indicate a suitable method of treatment, and as such this attempt to elucidate the history of the European fauna should be received.

This work was written as the outcome of a paper published in the *Proceedings of the Royal Irish Academy* (3rd series, vol. iv., 1897), "On the Origin of the European Fauna." A summary of that paper appeared in *Nature* (vol. lvi., 1897), and fuller extracts of more important parts, with some criticisms, in the *Geological Magazine* (N.S., sec. iv., vol. iv., 1897). I freely acknowledge the value of these criticisms, which have largely assisted me to amplify and to improve upon the ideas laid down in the paper.

I have found that it greatly facilitates comprehension of the arguments used, to give a few maps indicating in a general way the extent of former seas and continents. I may in this way, as Mr. Kendall has pointed out, have submerged many square miles of land which had never been covered by the sea,—at least not within recent geological times,—but the maps were intended as illustrations of my views in a broad spirit only.

Some zoologists may be surprised that, in some cases, I have not followed the latest views in revised nomenclature. I felt that in a work of this kind it was of supreme importance to employ names still current in our leading text-books, such as *Lepus variabilis* for the Mountain Hare, instead of *Lepus timidus*. After each chapter I have endeavoured to give a short summary of contents, while a bibliography of the principal works and papers consulted will be found at the end. I should also acknowledge the aid which I have received from such excellent works of reference as the British Museum Catalogues of Birds, by Dr. Bowdler Sharpe, and those of Reptiles, Amphibia, and Fishes, by Dr. Günther and Dr. Boulenger. The valuable works on Mammalia by Sir W. Flower, Mr. Lydekker, Mr. Grevé, and Dr. Trouessart, were indispensable to me.

To Sir William Flower, Mr. Lydekker, Professor Sars, and Professor Smitt, I am especially indebted for allowing me to reproduce drawings from their works, and to my friend Mr. Welch for some beautiful photographs. The Council of the Royal Irish Academy also kindly gave me permission to reprint the maps used in illustration of my paper. Professor Haddon first suggested my writing this book, and gave me many useful hints; and great assistance was rendered me by my colleague, Mr. G. H. Carpenter, in revising the proofs. To both of these kind friends I desire to acknowledge my deep sense of gratitude.

R. F. SCHARFF.

THE HISTORY OF THE EUROPEAN FAUNA.

CHAPTER I.

INTRODUCTION.

EVERY student of natural history, whether he be interested in birds, butterflies, or shells, contributes his share of facts which help to show how the fauna of his country has originated. The capture of a Swallow-tail or of a Marbled White Butterfly in England at once furnishes material for reflection as to the reason of its absence from Scotland and Ireland. Why should the Nightingale allow its beautiful song to be heard in England, and never stray across the Channel to the sister isle or cross the borders of North Britain? Lovers of bird-life and sportsmen, who have observed the habits of the Ptarmigan in the wild mountain recesses of Scotland, are aware that nowhere else in the British Islands do we meet with this interesting member of the grouse family, and many no doubt have allowed their minds to dwell upon the causes of its singularly local distribution.

I

All these animals have a wide range in other parts of the world. In past times, before man began to make observations on the geographical distribution of birds and butterflies, or even before the appearance of man in Northern Europe, they may have lived all over the British Islands. For some reason or other they are perhaps dying out or withdrawing towards their original home, which may either be northward, or to the east or south. If we had some clue as 'to their former history from fossil evidence—or, in other words, if their remains had been preserved to us in geological deposits,—we should have less difficulty in deciding this problem. But butterflies are scarcely ever preserved in a fossil state, and birds very rarely. We know little or nothing, therefore, of their past history from direct evidence, and are obliged to trust to indirect methods of research which will be indicated later on.

Mammals and Snails tell us their story more plainly. The bones of the former and the shells of snails are easily preserved, and thus furnish us with the necessary data as to their past history, for we find them abundantly in most of the recent geological deposits. Among the mammals of the British Islands there are some instances of distribution which much resemble those I have quoted. Thus the Arctic Hare (*Lepus variabilis*) is in the British Islands confined to Ireland and to the mountains of Scotland; and if it were not for the fact that its bones have been discovered in a cave in the south-west of England, we should perhaps never have known that, formerly, it must have inhabited that country as well. Of other mammals we possess fossil and also historical evidence of their having once lived in these islands. Such are the Wolf and the Wild Boar, both of which were abundant in Great Britain and Ireland. The latter is a distinctly southern species. We assume this, because its remains have never been found in high northern latitudes; nor does it now occur in Northern Europe or Northern Asia, whilst all its nearest relatives live in sub-tropical or tropical climates. The Arctic Hare, on the contrary, has probably come to us from the north. Its remains are unknown even in Southern Europe, and the more we approach the Arctic Regions, the more abundant it becomes. Thus we have here two instances of British mammals, one of which, the Wild Boar, has died out—as it were in a southerly direction; whilst the other, the Arctic Hare, is apparently retreating towards the north.

There are also some British mammals of which we have no fossil history, at least of which no remains have as yet been found in these islands. Such a one is the Harvest Mouse (*Mus minutus*). It has a somewhat restricted range in England, and only just crosses the Scottish border in the east. From the rest of Scotland and from the whole of Ireland it is absent. To judge from this distribution, in connection with the fact of its being unknown as a British fossil species, it is probably a late immigrant

to England, and has not had time to spread, throughout Scotland at any rate. But it is also absent
from Scandinavia, from the Spanish peninsula, from
almost the whole of Italy and the Alps, as also from
the Mediterranean Islands, whilst the little mouse
occurs abundantly right across Siberia. We shall
learn more about centres of dispersion later on;
meanwhile I should mention that such a distribution
indicates that the Harvest Mouse has most likely
originated in the east, and has spread from there
westward in recent geological times.

Conchologists have long ago been acquainted with
the fact that many molluscs, for example the so-called
"Stone-cutter" Snail (*Helix lapicida*) and the "Cheese
Snail" (*Helix obvoluta*), have a very restricted range
in the British Islands. Both are entirely absent
from Scotland and Ireland, the Cheese Snail being
confined to South-eastern England. The Stone-
cutter has rather a wider range, is even known from
a Welsh locality, and is met with as far north as
Yorkshire. Their distribution would indicate, therefore, that while both are recent immigrants, the
Cheese Snail is probably the last comer. This
supposition is in so far supported by fossil evidence,
as the latter is unknown in the fossil state, whilst the
Stone-cutter has been described by Messrs. Kennard
and Woodward (p. 243)[1] as occurring in the cave
deposit known as the Ichtham fissure, and also from

---

[1] The numbers in brackets throughout this work refer to the page-
number in the Bibliography at the end.

several English pleistocene and holocene deposits. The Stone-cutter can scarcely be looked upon as a very recent immigrant in the light of this evidence, though we have no proof of its having ever had a much wider range in the British Islands than it has to-day.

Among the lichens, which so abundantly cover the rocks and trees in South-western Ireland, and which impart such a characteristic feature to the scenery, we find a beautifully spotted slug (*Geomalacus maculosus*).[1] It is a stranger to the rest of the British Islands, and indeed occurs nowhere else in Northern Europe. We have to travel as far as Northern Portugal before we again meet with it, and it is there also that its nearest relations live.

Many more similar examples might be quoted, but enough, I think, has been said to show that the British fauna is made up of several elements whose original homes may lie widely apart and in different directions. We have fossil evidence that some of the northern species, and also a few of the southern ones, have become extinct within comparatively recent times; others are apparently on the verge of extinction, whilst many not only maintain their position in the constant struggle for existence, but are even extending their range.

The problem of tracing the origin of the British fauna, or at least that of some of the more characteristic members of every section or element,

---

[1] A map giving its exact distribution in Ireland will be found on p. 300, and a figure of the slug on p. 298.

appears at first a somewhat difficult task. Indeed, the means of dispersal of the various groups of animals are so different that it occurred to me it might be better to deal with the mammals, the birds, the reptiles, and so forth, all separately. This idea I have attempted to follow to some extent, with most satisfactory results. The British fauna of the present day is no doubt complex, but no more so than the fauna of the most recent of our geological deposits—the Pleistocene. However, when we go back still further and look at the earlier Tertiary remains, we find the fauna becoming less complex. Northern species disappear, and the strata are entirely filled with the remains of southern animals and plants. Geologists indeed are quite unanimous in their belief, that the fauna of the British Islands during the earlier epochs of the Tertiary Era was a southern one; that it then gradually became more temperate, until at last, in more recent times, decidedly northern forms invaded the country. These seem to have driven out—to some extent at least—the southern species; but more recently again, the southerners, reinforced by an eastern contingent, appear to have gained territory and are advancing into the area held by the northerners. The eastern invasion does not seem to have affected Ireland at all, and we find the country there divided between the southern and northern animals. We can thus roughly construct a map as I have done here, showing, by means of horizontal and sloping lines, the principal areas inhabited at the present time by the species of northern, southern, and eastern origin (Fig. 1).

FIG. 1.—Map of the British Islands, indicating approximately the areas inhabited by the northern, southern, and eastern animals. The horizontal lines represent the areas of northern species, the sloping lines those of southern and eastern ones.

In the problems which are being discussed in this work I have often found it of advantage, in order to facilitate the comprehension of the arguments used, to give maps. Some of these represent the geographical conditions at the particular epoch referred to in the text, but they merely claim to give a general idea. There was never any intention to make them correspond with all the data of which we have geological evidence. They are what I might call "diagrammatic." In comparing them with reconstructions of former physical geography such as have been attempted from time to time, I hope geologists will therefore deal leniently with the faults I may have committed, and remember that the maps are "impressions," or "diagrams," and not faithful representations of all the geographical revolutions witnessed by some of our remote forefathers at any particular period.

The knowledge we gain from a study of the British Tertiary deposits enables us to affirm positively that both the eastern and the northern species arrived in these islands comparatively recently, but that the southern forms must have migrated northward from the Continent long ages ago. Since the northern and the eastern migrations—that is to say, those coming from the north and east—were the last to arrive in Northern Europe, the remains of the animals contained in the most recent deposits of that portion of our continent will furnish us with a clue as to the extent of the area inhabited by them. This is not all, however. It is also possible to discover from these remains the direction which the animals that they belonged to came from. As we shall learn later on, a migration on a vast scale entered Europe during the Pleistocene epoch—the most recent of the geological epochs, during which great extensions of glaciers occurred in the mountainous regions of Europe. The latter period is known to us as the Ice Age or Glacial period. This will be described more fully in Chapter II., meanwhile I may mention that we presume that this migration came from the east, because no remains of the members of that particular fauna are known from Spain, Southern Italy, Scandinavia, Ireland, or from the Balkan peninsula. The number of species evidently belonging to this same migration, moreover, become fewer as we proceed westward, and a large proportion of them still inhabit Northern Asia, though most of them are now extinct in Europe. After having thoroughly studied such a recent geological migration, we learn to understand others better, though the more ancient they are, the fewer are the traces and the more difficult are they to follow.

Then again we have to take into consideration the fact, that whilst mammals, particularly the larger herbivores, are forced to migrate frequently owing to scarcity of food or temporary changes of climate, many of the invertebrates remain practically unaffected by either. Most of our land mollusca, for instance, are satisfied with meagre provender, and stand extremes of climate well, as long as there is sufficient moisture. As a result of their peculiar disposition, many of them, no doubt, have survived through

several geological epochs, and have witnessed vast
geographical revolutions in their immediate surroundings, whilst mammals are comparatively short-lived.
Being driven from one country to another, and exposed to innumerable enemies, new types appear and
old ones rapidly vanish ; in fact, there are almost
constant changes in the mammalian fauna as we
pass from one epoch to another.

I have until now referred more particularly to the
British fauna and the North European in general,
because the history of our own animals interests
us all more than those of any other European area.
It is, moreover, preferable to commence our investigations into the origin of the European fauna
by the study of a small district. This should, if
possible, be an island. If we took a slice of the
continent like France or Germany, we should find
the problem more complex. Instead of choosing
the British Islands, we might, however, take an island
like Corsica or Sardinia. In either of these we should
discover peculiarities in the composition of their fauna
precisely similar to those which I have indicated to
be present in the British fauna. We should find
probably a more striking endemic[1] element, which

---

[1] The term *endemic* will be employed throughout this work as applied
to species peculiar to a country and not found elsewhere. *Autochthonous* will be used in speaking of a species which has originated in a
country to which, however, it is not peculiar; *e.g.*, the Chamois is an
autochthonous Alpine species, but occurs also in the Pyrenees and
Caucasus. An *indigenous* species is one native to a country, as

with us is so meagre that it can almost be left unnoticed; the main features, however, remain nearly the same. The fauna of both of these islands is composed of a strong southern element, of an eastern and a northern one, and in addition we have here species whose ancestors lived in Western Europe.

Before investigating more minutely the problems suggested by the composition of the faunas of these insular and also of some continental areas, it is necessary that we should thoroughly understand all about the migrations of animals. One of the principal objects of this work is to show how the autochthonous animals of Europe, *i.e.*, those which have originated there, may be distinguished from the immigrants, and to trace the latter to the home of their ancestors. But in doing so, it is necessary to refer to the many important geographical changes which have occurred in Europe during the latest geological epochs. The study of the geographical distribution of the European fauna, as expounded in this work, will in many instances confirm the theories as to geographical changes based upon geological foundations. But in every case the views herein advocated are founded upon the geographical distribution of living and extinct organisms alone.

A terrestrial mammal like the deer can, under ordinary circumstances, only reach one part of a country from another by walking or running to it; but a beetle, such as the cockchafer, has two different

modes of progression. It may walk or fly. In both,
however, there is a third mode of transport—an in-
voluntary one. The deer may be suddenly seized by
a flood whilst crossing a river, and carried far away
without necessarily coming to grief. The beetle in
a similar manner could be transported to a distant
country, or it might be caught in a whirlwind and
blown hundreds of miles off.

We may thus distinguish between the natural or
active and the accidental or passive means of distribution of animals. The active mode of dispersal again
may be only migratory, as in most animals, or periodic
and migratory, as in some birds and fishes. It is of
course the tendency of every species to spread in all
directions from its original home, provided it does not
encounter obstacles, such as want of food, unsuita-
bility of climate or soil, or barriers such as mountains,
rivers, or the sea. Birds might be thought to be little
interfered with by any of these barriers, but, as Dr.
Wallace has shown, they are almost as much affected
by them in their distribution as mammals are.

This then is the ordinary migratory distribution.
Periodic distribution obtains with migratory birds
and fishes. The annual flight of swallows to their
northern summer residence comes under the heading
of periodic migration or distribution, but apart from
this, the swallow must seek to extend its range by
the ordinary method, like every other animal.
Similarly, the herring migrates periodically into
shallow water to spawn, only to return again to its

deeper home, where, as its numbers increase, there must be a tendency to spread. We have in these cases, therefore, both a periodic and an ordinary movement of migration.

Now, in studying the composition of a fauna, and especially its origin, it is of the utmost importance to be able to determine approximately the percentage of accidental arrivals and of the ordinary migrants— that is to say, of those which have reached the country owing to accidental distribution, and of animals which have adopted the usual course of migration. It is of all the more import to review this subject of accidental, or, as Darwin called it, "the occasional means" of distribution, as both he and Dr. Wallace have, I venture to think, somewhat over-estimated its significance. No one doubts that accidental transportal takes place, but the question is whether the accidentally transported animals arrive living and reach a spot where suitable food is procurable, and whether they are able to propagate their own species in the new locality. For it must be clear to anybody that the accidental transportal of a beetle or of a snail to a new country cannot affect its fauna or add one permanent member to it unless all these conditions are fulfilled. As a matter of fact, only exceedingly few instances are on record of man having witnessed, for example, the accidental transportal across the sea to an island of a live animal.

To mention an example, Colonel Feilden informs

us (*Zoologist*, 1888) that, when living on the island
of Barbadoes, an alligator arrived one day on the
shore, and at the same time a tree measuring 40 feet
in length, which was recognised as a Demerara species,
was likewise stranded. He thinks that there can be
no doubt that the alligator, which was alive when it
reached Barbadoes, was transported by the tree, thus
covering a distance of 250 miles from the nearest
land. Numerous observations on the accidental
transportal of seeds and tree-trunks from one island
to another, and from a continent to an island, have
been recorded, and even on our own shores we may
witness the occasional arrival of such vegetable products from a far distant land. On the west coast of
Ireland it not unfrequently happens that large West
Indian beans are stranded, and in this as well as in
many other similar cases the seeds have often proved
none the worse for their prolonged immersion in sea-
water. That locusts are sometimes blown to great
distances from the land is not so surprising, since
their power of steering through the air is very limited.
Darwin mentions (p. 327) having caught one 370
miles from the coast of Africa, and that swarms of
them sometimes visited Madeira. Sir Charles Lyell
relates that green rafts composed of canes and brush-
wood are occasionally carried down the Parana River
in South America by inundations, bearing on them
the tiger, cayman, squirrels, and other quadrupeds.

But though actual observations of such abnormal
instances of the dispersal of animals are few, many

experiments have been made to demonstrate the possibility of a passive transportal of species over wide distances. It was especially Darwin who gave a great stimulus by setting the example to those interested in natural history in the conduct of such researches. He was struck by the fact that, though landshells and their eggs are easily killed by sea-water, almost all oceanic islands, even the smallest and most isolated, are inhabited by them, and felt that there must be some unknown but occasionally efficient means for their transportal (p. 353). To quote his words: "It occurred to me that land-shells, when hibernating and having a membranous diaphragm over the mouth of the shell, might be floated in chinks of drifted timber across moderately wide arms of the sea. And I find that several species in this state withstand uninjured an immersion in sea-water during seven days: one shell, the *Helix pomatia*, after having been thus treated and again hibernating, was put into sea-water for twenty days, and perfectly recovered. During this length of time the shell might have been carried by a marine current of average swiftness to a distance of 660 geographical miles. As this *Helix* has a thick calcareous operculum, I removed it, and when it had formed a new membranous one, I again immersed it for fourteen days in sea-water, and again it recovered and crawled away. Baron Aucapitaine has since tried similar experiments: he placed one hundred land-shells, belonging to ten species, in a box pierced with holes, and immersed it for a fortnight in the sea. Out of the hundred shells, twenty-seven recovered. The presence of an operculum seems to have been of importance, as out of twelve specimens of *Cyclostoma elegans* which it thus furnished, eleven revived. It is remarkable, seeing how well the *Helix pomatia* resisted with me the salt-water, that not one of fifty-four specimens belonging to four other species of *Helix* tried by Aucapitaine, recovered. It is, however, not at all probable that land-shells have often been thus transported; the feet of birds offer a more probable method."

We have here positive evidence that such shells as *Helix pomatia* and *Cyclostoma elegans* might easily be transported to an island from the mainland. The former occurs in France, Holland, and England, and the latter all along western continental Europe and England. And yet neither of these species inhabits the Canary Islands, Madeira, or Ireland, none of which are at too great a distance from Europe to be within easy reach for a floating object. The fact that *Cyclostoma elegans* does not live in Ireland is of particular interest in connection with the floating-theory just quoted, as on all sides of Ireland dead specimens have been picked up on the shore, showing that marine currents carry specimens and have thus transported them for countless centuries. Nevertheless the species has not established itself in Ireland. If such a fate meets a land-shell of the type of *Cyclostoma elegans*, it may be asked, with some justification, what chance slugs or the smaller non-operculated species would have to

reach an island like Ireland alive from the mainland, and to colonise it successfully.

Both slugs and their eggs are killed by a short
immersion in sea-water, as I have proved experi-
mentally. I have also subjected slugs, in the act of
crawling on twigs, to an artificial spray of sea-water.
This seemed to irritate their tender skins to such an
extent that they curled themselves up, released their
hold on the twig and let themselves drop to the
ground. If we supposed, therefore, that a slug had
successfully reached the sea, transported on a tree-
trunk, the moisture would tend to lure it forth from
its hiding-place under the bark, whilst the mere spray
would prove fatal to its existence. Those species of
snails and slugs which lead an underground existence,
rarely venturing above ground, such as *Testacella* and
Coccilianella, would have even less chance of being
accidentally carried to some distant island.

The suggestion advanced by Darwin (p. 353), that
young snails just hatched might sometimes adhere to
the feet of birds roosting on the ground and thus be
transported, appears to me so extremely improbable
as to be scarcely worth serious consideration. Indeed,
as Darwin himself acknowledged later on, it does not
help us very much to suggest possible modes of
transport. What we require is direct evidence. How
far we are, however, from obtaining it, may be inferred
from Mr. Kew's remark (p. 119), that "we have
little or no actual evidence of precise modes of
dispersal even for short distances on land."

2

. A very curious statement was made by a well-known French conchologist, the late M. Bourguignat, with regard to introductions of mollusca. Whether he had any actual facts collected in support of it, I cannot say, but he maintained that species accidentally transported, with the exception of those under maritime influence, can only be acclimatised from north to south, and not from south to north—from east to west, but not from west to east (p. 353).

The whole theory of the accidental or abnormal dispersal of mollusca appears to have been originated by Darwin, in order to account for their presence on so-called *Oceanic islands*. His views were strongly supported by Wallace, who defines these islands (p. 243) as those which are of volcanic or coralline formation usually far from continents, entirely without indigenous land mammals or amphibians, but with a fair number of birds and insects, and usually with some reptiles.

I do not wish it to be understood that I am in any way undervaluing the great works of these distinguished naturalists. Darwin's views have had more influence in advancing Zoology than those of any man, and his fame is unassailable. Nevertheless, I feel that his theories regarding the origin of the faunas of oceanic islands require revision.

The formerly prevalent belief of the permanence of ocean basins has been shaken by the utterances of some of the greatest geologists of our day, whilst many positively assert that what is now deep sea of more than 1000 fathoms was dry land within comparatively recent geological epochs. Thus the Azores are classed by Darwin and Wallace among the oceanic islands—that is to say, among such as have received their fauna and flora by flotsam and jetsam. But Professor Neumayr believes, on geological grounds, that the Old and New Worlds were connected by a land-bridge during Tertiary times right across the Atlantic, and that the Canary Islands, Madeira, and Azores (p. 547) are the last remnants of this continent. This meets with the entire approbation of Dr. von Ihering, who has recently re-investigated the subject from a faunistic point of view (p. 135). Take another instance of one of Wallace's most typical oceanic islands, the Galapagos Group. Their fauna and flora have recently been most thoroughly re-explored by an American expedition, the result of which, according to Dr Baur, goes to show that these islands must have formed part of the mainland of South America at no distant date. The fauna and flora are therefore to be regarded as having reached them in the normal mode, viz., by migration on land. According to Mr. Beddard (p. 138), it is difficult to see how earthworms could be transported across the sea. Floating tree-trunks have been observed far out at sea, but unless the water remained absolutely calm during the long period necessary for the drifting by currents so that no splashing occurred, the worms would probably be killed. Yet earthworms do occur on oceanic islands. It is indeed quite possible that our views with regard to the origin of the remainder of the Pacific Islands may change very materially, and once more revert to what Dr. Gould expressed nearly fifty years ago in the following words : "From a consideration of the land-shells on the Pacific Islands, it seems possible to draw some fair inferences as to the relations of the lands which once occupied the area of the Pacific Ocean, and whose mountain peaks evidently now indicate or constitute the islands with which it is now studded." Indeed Dr. von Ihering goes so far as to positively state that in his opinion the Polynesian Islands are not volcanic eruptions of the sea floor, which being without life were successively peopled from Australia and the neighbouring islands, but the remains of a great Pacific continent, which was in early mesozoic times connected with other continental land masses (*a*, p. 425).

Before coming to a decision on the part played by flotsam and jetsam in the constitution of an island fauna, those who have studied the problem on the spot should, however, have a voice in the matter. And though, from my experience in northern latitudes, I feel sure that island faunas there are but slightly affected by occasional dispersal of species, Mr. Hedley, who has made the fauna of the Pacific Islands his special study, assures me that drift migration plays an important *rôle* in that region. I hope we may soon have a more detailed account of his particular observation bearing on this interesting subject.

On the other hand, Mr. Simpson, who has gained considerable experience of oceanic dispersal in the West Indian region, though he acknowledges having often noticed bamboo rafts, which would be suitable in the transportal of invertebrates, nevertheless does not attach much importance to this means of distribution. "The fact," he remarks, "that the operculates (operculate land-shells) form so large a proportion of the Antillean land-snail fauna, that a majority of the genera are found on two or more of the islands and the mainland, while nearly every species is absolutely restricted to a single island, appears to me to be very strong testimony in favour of a former general land connection" (p. 428).

Amphibians are affected in the same manner by sea-water as slugs are. The accidental transportal of an amphibian from the mainland to an island is therefore almost inconceivable. And the presence of frogs, toads, and newts in the British Islands, in Corsica and Sardinia, indicates, if nothing else did, that all these islands were at no distant date united with the continent of Europe.

As regards the terrestrial reptiles, the case is somewhat different. Many of them readily take to the sea, and, as probably all snakes and some lizards are able to swim, it is possible that sometimes, though very rarely, they might reach islands if not too far from a continent. Instances of accidental transportal of land-reptiles to islands have actually been observed. But the fact of the occurrence of such instances by no means proves that reptiles thus conveyed are able to establish themselves permanently in their new home. Sir Charles Lyell records in his *Principles of Geology* that a large boa-constrictor was once seen floating to the island of St Vincent, twisted round the trunk of a tree. It appeared so little injured by its long voyage from South America, that it captured some sheep before it was killed.

Mammals might be accidentally conveyed to islands on such rafts as have been described by Sir Charles Lyell, and there are instances on record of their having crossed short distances of sea by swimming. Elephants and also deer and pigs are good swimmers, the former having been known to swim for six hours at a stretch. "But," remarks Mr. Lydekker (p. 13), "it may be assumed that about twenty miles is the utmost limit which mammals are likely to cross by swimming, even when favoured by currents. Such passages as these must, however, be of very rare occurrence, for a terrestrial mammal is not likely to take it into its head to swim straight out to sea in an unknown direction. Moreover, supposing a mammal, near to a particular island, to have arrived there by swimming, unless it happen to be a pregnant female, or unless another individual of the same species but of the opposite sex should arrive soon after (a most unlikely event), it would in due course die without being able to propagate its kind."

All zoologists, indeed, are quite in accord with Dr. Wallace's view as expressed in *Island Life* (p. 74). "Whenever we find that a considerable number of the mammals of two countries exhibit distinct marks of relationship, we may be sure that an actual land-connection, or at all events an approach to within a very few miles of each other, has at one time existed." As all the European islands come under this category, their mammals exhibiting distinct relationship with those on the European continent, they all have been connected with it formerly.

Perhaps the most powerful of all agents in the transportal of species by accidental means is man. But his actions may be accidental as well as intentional. We have therefore to distinguish between the animals disseminated all over the world by pure chance, and those which have been introduced into new countries purposely. Invertebrates, such as snails, centipedes, woodlice, beetles, and cockroaches, are constantly being unintentionally carried with vegetables, fruit, trees, and with timber from one country to another. Earthworms are sometimes transported in the balls of earth in which the roots of trees are enveloped. As regards molluscs, Mr. Kew believes (p. 178) that during the last three centuries at least, human agency has influenced their disposal more than all other causes taken together. A large number of species of invertebrates in America are said to owe their existence in that country to accidental introduction by man. In

most cases, however, no particular reason can be
assigned why they should have been thus introduced,
and as a matter of fact there are always individual
differences of opinion as to the precise number of
such.   Certain it is, that though the number of
supposed introductions from Europe to America
is very large, those which have been carried from
America to Europe is exceedingly small.   In fact,
I remember only two instances of accidental animal
importations from America which have firmly
established themselves in Europe, viz., a small
fresh-water mollusc, *Planorbis dilatatus*, and the
much-dreaded vine-pest, *Phylloxera vastatrix*.

As a rule the animals die out very shortly after
their arrival on foreign soil.   Many instances,
nevertheless, are on record, especially in the case
of molluscs, where snails thus transported have not
only survived but are apparently in a flourishing
condition and spreading.   *Helix aspersa*, for example,
our large garden snail, has been naturalised in many
foreign countries by French and Portuguese sailors,
who had taken them on board their ships as food.

It certainly cannot be denied that a number of
species among almost all groups of invertebrates have
been unintentionally conveyed by man from Europe
into foreign countries.   It has been proposed by Dr.
von Ihering to apply the term "cenocosmic" to those
species which have become spread all over the world
through artificial means, and thus to distinguish
them from cosmopolitan ones which have attained a

similar range naturally. The latter he calls "palin-cosmic" species (*a*, p. 422). Many so-called ceno-cosmic ants are believed by Dr. von Ihering to be palincosmic. We are altogether too apt to regard cosmopolitan as synonymous with introduced, and we should hesitate before concluding that because one of our common European species occurs in Australia or South America, it must have been transported there recently by human agency. Some of our widely-distributed forms are probably of very great antiquity, and may have spread to distant lands in early Tertiary times, when a different state of the geographical conditions enabled them to do so.

I cannot quote a more appropriate instance than the molluscan fauna of Madeira. No less than thirteen of the Madeiran snails are looked upon as having been introduced from Europe by human agency, on the sole evidence that these happen to be common European species. Yet the correctness of this supposition must be questioned in face of the interesting observation made by Darwin (p. 357), "that Madeira and the ad-joining islet of Porto Santo possess many distinct but representative species of land-shells, some of which live in crevices of stone ; and although large quantities of stone are annually transported from Porto Santo to Madeira, yet this latter island has not become colonised by the Porto Santo species. Never-theless, both islands have been colonised by European land-shells, which no doubt had some advantage over the indigenous species." Darwin, therefore, meets the evident anomaly by suggesting that the European species are supposed to possess some advantages as colonisers. But the true explanation appears to me to lie in the supposition that the European land-shells found in the Madeiran Islands are all, or for the greater part, ancient forms which survived both there and on the continent, whilst the remainder of the forms inhabiting these islands are either such as are now extinct in Europe, or have become modified since their arrival there from the continent at a time when extensive land-connections allowed a free migration by land.

The theory of accidental introductions is an extremely popular one. It allows free scope to a host of speculations, and once the idea has taken firm root that a certain species is introduced, especially among the class of naturalists who by way of experiment are wont to create new centres of dispersion in their own neighbourhood, evidence to the contrary must be of the most convincing nature to shake the popular belief. Thus, it is almost regarded as an established fact by conchologists and others, that the fresh-water mussel (*Dreyssensia polymorpha*) was introduced into England at the beginning of this century. Though it has been proved that this species is quite unable to live in pure sea-water, yet the view that it has been carried from the Black Sea ports to this country attached to the bottom of ships is maintained by many, whilst others incline to the theory that the shell came with timber. But *Dreyssensia polymorpha* was by no means always confined to the Caspian and Black Sea areas; it occurs abundantly in the lower continental boulder-clay (see p. 230), and no doubt it had at one time a much wider geographical distribution. It appears to me possible, that it was able to maintain itself in certain fresh-water lakes and slow-flowing rivers in Northern Europe, from which it might have spread since the introduction of canals into Europe at the beginning of the century. As the larva of this fresh-water mussel is free-swimming, its propagation is much favoured by canals. Quickly-flowing rivers are fatal to its existence, since the delicate larvæ are swept out to sea and perish. Such an hypothesis as this is strengthened by the fact of its recent discovery in a sandy layer fifteen feet below the present surface under the streets of London in a deposit which probably, as Mr. Woodward remarks (p. 8), was accumulated in the early days of the city's existence. In spite of Mr. Woodward's interesting find, and Dr. Jeffreys' opinion, who always maintained that this shell was indigenous to England, popular belief still clings tenaciously to the introduction theory.

Among man's intentional introductions into a new country, no instance is better known than that of the rabbit to Australia. Rabbits are entirely confined to Europe. In their transplantation to Australia they were carried to a country with a different climate and among new surroundings. Yet the rabbits flourished, and within comparatively few years increased to such

·an extent as to become a burden and pest to the country. It may be remembered though, that, owing to the complete absence of small carnivores, which act with us as a check upon the too rapid increase of this rodent, the speed with which it established itself in the new surroundings is not so very surprising.

Many of the English settlers in the New World felt that America lacked the presence of our familiar birds. The homely sparrow was therefore brought over, with the result that the Agricultural Department of the United States is now devising means for its destruction, so rapid has been its increase.

Similarly, the inhabitants of Jamaica, annoyed by the great profusion of rats in their island, sent over to India for a number of mongoose. These have decimated the rats since their arrival, but they have multiplied to such an extent as to be a serious menace to the native fauna.

To give an instance nearer home, the Capercaillie (*Tetrao urogallus*) was successfully introduced into Scotland in 1837. From its different centres of distribution it is spreading in all directions where sufficient cover is obtainable. But this case differs from the others very materially, in so far as this bird was formerly a native of Scotland, and only became extinct during the last century.

However, although there are many examples of undoubtedly successful introductions by human agency, quite as many, or perhaps more, unsuccessful ones might be quoted. In fact, it is by no means easy to establish a species in any new locality. Frequently it happens that the species seems to be on the increase at first, but then there is a decline, and after a few years the new plantation has entirely vanished. In other cases, the species disappears immediately after the introduction takes place, or lingers on for many years if it receives special and uninterrupted protection.

It may not be generally known that the English Hare (*Lepus Europæus*) is not found in Ireland, where the Mountain Hare (*Lepus variabilis*) alone occurs. Attempts to acclimatise the English species have been made in a number of places in Ireland, but many of them have been failures, and not one of them has been a signal success.[1] Similarly, the endeavour to introduce the French or Red-legged Partridge (*Caccabis rufa*) into Ireland has met with a like result. According to Dr. Day, it was tried during the summer of 1869 to naturalise the Sterlet (*Acipenser ruthenus*) from Russian waters into the Duke of Sutherland's River Fleet by importing artificially impregnated ova. From one hundred and fifty to two hundred lively young sterlets are said to have been turned out, but nevertheless the experiment met with no success. Several fortunately abortive efforts were also made in British rivers to establish *Silurus glanis*, a hideous monster of a fish, and quite unpalatable.

---

[1] I might refer any one more specially interested in these introductions to an article on this subject in the *Irish Naturalist* of March 1898, by Mr. Barrett-Hamilton.

The Natterjack Toad (*Bufo calamita*) has a very local distribution in the British Islands. In Ireland it is found only along the coast of Dingle Bay in County Kerry, where it is known among the peasantry as the Black Frog. There is no doubt about its being indigenous there, and though it has not spread beyond the very limited area of its habitat, the Irish climate cannot be said to be unsuited to its existence. Yet it seems to be extremely difficult to acclimatise it elsewhere, for though no less than sixty specimens were turned out in Phœnix Park, Dublin, about forty years ago, none of them were ever seen afterwards. They were placed in the vicinity of one of the lakes, so as to give them ample scope for breeding and developing the young, and in surroundings which were considered eminently suitable at the time.

It has occasionally happened, too, that animals are introduced by kindly-disposed persons with the view of adding a species to their fauna, in complete ignorance of their previous existence in the country where they wished to naturalise them. Thus we are told that in the year 1699 one of the Fellows of Trinity College, Dublin, procured Frog's spawn from England in order to add that amphibian to the Irish fauna. It was placed in a ditch in the College Park, whence the species is supposed to have gradually spread all over the island. This story is quoted by many writers as the true history of the Frog in Ireland, and is given as an example of the rapidity with which animals spread. Unfortunately the would-be introducer seemed unaware that, according to Stuart's *History of Armagh*, the first Frog which was ever seen in Ireland made its appearance in a pasture field near Waterford about the year 1630, that is to say, seventy years before its introduction in Dublin.[1] But even Stuart was mistaken in supposing that no Frog had ever been seen in Ireland before, since Giraldus Cambrensis, in his *Topography of Ireland*, mentions that a Frog was found in a meadow near Waterford in the year 1187.

Certain British species of vertebrates are generally looked upon as introduced species, though we cannot trace any record of their first establishment, and it is quite possible that, though there was local extinction and subsequent local re-introduction, they are truly indigenous and may never have become totally extinct. Such are, for instance, the Rabbit (*Lepus cuniculus*) and the Pheasant (*Phasianus colchicus*). The latter certainly had become naturalised in England before the Norman invasion.

But cases of introduction such as those above referred to are by no means confined to the vertebrates, similar instances among invertebrates being numerous enough. I am sure every naturalist is personally acquainted with a good number, and it is hardly necessary that I should quote in any detail after

---

[1] I should recommend those who are particularly interested in the full history of the Irish frog to read the notes on this subject contained in vol. ii. of the *Irish Naturalist*.

what has been said on the subject generally. The two species of snails, *Helix pomatia* and *Cyclostoma elegans*, both of which occur in England, and which I had occasion to mention among those experimented on by Darwin, were turned out in several suitable localities in Ireland by Thompson, but failed to establish themselves. The former, according to Mr. Kew, was also introduced into Scotland and Norway, whilst fifty or sixty specimens were brought to Petersfield in England, but none of these trials at acclimatisation were successful. As among vertebrates, a large number of the so-called successful introductions rest upon insufficient evidence.

When we once more carefully review the evidence as to the undoubted difficulty attendant on intentional introduction of animals by human agency, placed as they often were in most suitable localities, we must feel that accidental introduction cannot play an important rôle in the making of the fauna of any country. Especially is this the case with an island fauna. Vertebrates are almost altogether excluded, and invertebrates must arrive singly as a rule, often stranded on an inhospitable and unsuitable shore. Their chances of surviving a passage by sea, of finding suitable food and shelter and a mate in order to procreate their species, appear to me infinitesimally small. Yet there may be some such cases. However, I quite agree with Mr. Andrew Murray—a high authority on geographical distribution—that "colonisation or occasional dispersal is insufficient to account for the character of the faunas and floras of oceanic islands; and I believe that the normal mode in which islands have been peopled, has been by direct continuity with the land at some former period, or by contiguity so close as to be equivalent to junction" (p. 15). "That a slight intermixture," he continues, "due to Mr. Darwin's colonisation, occurs in many (probably in all) I am ready to admit; and from instances to be afterwards noticed, I am disposed to reckon the proportions of such intermixtures in the flora, in the most favourable circumstances, at not more than two per cent. In the fauna I think it must be much less."

Mr. Murray's views, though they relate only to oceanic islands, are likewise applicable to continental islands such as our own. I think we might take the admixture in the British fauna due to occasional, including human introduction, as amounting to five per cent. It is better to take a high estimate, so as to include all the species about whose native land there might be some reasonable doubt. Now of what importance, after all, is this five per cent.? The remaining ninety-five per cent. of the species of animals belonging to the British fauna undoubtedly migrated to these islands in the normal way by land.

It is of great importance, in dealing with the question of the origin of the British fauna, to thoroughly grasp this conclusion—*that ninety-five per cent. of the animals have reached us by land.* We can afford in fact to ignore the five per cent. altogether. It is an insignificant factor. As regards

·the botanical aspect of the question, botanists are
quite in accord with the zoologists, and entirely
share their views in the belief of a former land-
continuity between the British Islands and the
Continent. "It cannot be denied," says Professor
Blytt (p. 32), "that a plant of one or another
species may, in an exceptional case, migrate, without
human assistance, all at once, across large tracts
of land and sea, and that such migration, if operating
during geological periods, might introduce a number
of species even into distant oceanic islands; but
when the question is of whole communities of plants,
such as the above enumerated elements in our flora,
then such an accidental and sudden transport across
large tracts can only be conceived to be at all
probable in the case of Arctic plants carried by
drifting ice to a bare country without native flora ;
as to the other species, we must imagine that the
*migration during the gradual change of climate has
proceeded slowly and step by step across connected
tracts of country.* In that manner we may assume
that our country has in the course of time obtained
its present covering of plants. Each of the above-
named elements in our flora has doubtless its
corresponding element in our fauna. The fauna
and flora of a region stand in relation of complicated dependence to each other. The animals
live on the plants. The fecundation of the plants
takes place in a great degree by means of insects ;
their seeds are often scattered by resident birds and

quadrupeds. Everything indicates that *conveyance to small distances is the rule*, and that sudden and long migration is the exception."

The conviction which has been gained by zoologists and botanists, that the British Islands once formed part of the Continent, is based on the present British fauna and flora. The remains, however, of animals which used formerly to live in these countries, such as the Mammoth, the Irish Elk, the Cave Bear, and many others, tell us the same tale. They could not have peopled England by swimming across the Channel, or even by walking across solid ice, as has once been suggested. Nothing but a land-connection induced them to explore this country more closely, and finally to decide on settling there.

The origin of the British fauna will be discussed more in detail in the third chapter. The methods of investigation adopted, along with a general scheme of this book, will be found in the next.

The manner in which the origin of the fauna of any particular continental area can be traced is very similar to that adopted in the case of an island. Portions of the continent of Europe can be shown to have been islands in former times. Thus the Crimea, now a peninsula united to the mainland by the narrow isthmus of Perekop, must have been an island in comparatively recent times. The absence of a number of striking and familiar South Russian species of mammals and reptiles proves this to have been the case. It was probably long after the appearance of man, though before historic times, that these changes took place.

We shall learn in the subsequent chapters, that, by a careful study of the fauna and flora the fact can be established, not only of the former connection of an island with a continent, but also whether such union existed (geologically speaking) within recent or more remote times. The better the fauna is known, both recent and fossil, the more precisely can the period of connection be indicated, and its duration determined.

CHAPTER II.

I INTEND to give in this chapter a general outline of the subject which will be discussed in the subsequent ones. This will include a brief history of the great events, in recent geological times, which have modified the evolution of the European fauna by the influence which they have exerted on the course of the successive streams of migration.

The composition of the European fauna is the first item which will have to be taken into consideration. But not only must the existing species of animals be dealt with : the extinct ones, too, at least those which have lived in Europe during late Tertiary times, will be useful for our inquiries. A knowledge of the past faunas is a most important factor in tracing the original home of the European animals.

Where a species first originated, whether this was in one or several places, or, in other words, where it first had its home, cannot be determined with absolute certainty in the present state of our knowledge, but as a rule it can be indicated approximately with a fair amount of precision. In a few instances, species may possibly have had a dual origin. The majority of naturalists doubt that there are any such, but it seems to me, that almost the same forces may have acted in different localities on certain forms so as to produce, in very exceptional circumstances, similar species. The vast majority of animals, however, have no doubt originated in one locality; or, we might say, almost all species have but one home.

We may assume that every animal gradually extends its range by migration, as the result of the natural increase of the species necessitating a search for fresh feeding grounds. Every species thus tends to slowly take possession of all the habitable parts of the globe to which it has access. They would all naturally spread from their original homes in every direction, unless prevented by an impassable barrier. We have already learned that to all land animals, the sea acts as such a barrier. Mountains and rivers act also in a similar way, but not to the same extent. It is not difficult to understand also that a forest may be a formidable barrier to a typical inhabitant of the open country and *vice versâ*, whilst a desert is impassable to almost all terrestrial organisms. Some species are scarcely affected by climate, and flourish equally well in the tropics and in temperate or cold countries ; the majority, however, are greatly influenced by it. "No more striking illustration," remarks Merriam (p. 38), "could be desired of the potency of climate compared with the inefficiency of physical barriers, than is presented by the almost total dissimilarity of the North American Tropical and Sonoran Regions, though in direct contact, contrasted with the great similarity of the Boreal Regions of North America and Eurasia, now separated by broad oceans, though formerly united, doubtless, in the region of Behring Sea."

To return to the composition of the European fauna, we now know positively that a number of the mammals and birds inhabiting Central and Eastern Europe are of Siberian origin. How they came, and when, will form the subject for discussion in Chapter V. At present it will suffice to mention that in the superficial deposits belonging to the Pleistocene series of the North European plain have been discovered the remains of many typical members of the Siberian Steppe-fauna. Some of these, such as the Saiga-Antelope (*Saiga tartarica*), Fig. 2, still inhabit portions of Eastern Europe, whilst others have retreated to their native land. But it might be asked, how is it known that these species did not originate in Europe, and thence migrate to Siberia? Because if they had originated on our continent, they would have spread there. They would have invaded Northern and Southern Europe, and they would probably have left some remains in Spain, Italy, or Greece. They would also have left some of their relations in Europe; but all their nearest allies, too, are Asiatic. Moreover,—and this completes, I think, the proof of their Siberian origin,—the Pleisto-

FIG. 2.—The Saiga-Antelope (*Saiga tartarica*). (From Lydekker's *Royal Natural History*, vol. ii. p. 298.)

cene remains of these animals in Europe become less abundant, and the number of species likewise decreases, as we proceed from east to west. With these remains of Steppe animals are generally associated those of others, which we must also look upon as Siberian emigrants, such as the Pikas or tailless Hares belonging to the genus *Lagomys*, the pouched Marmots (*Spermophilus*), and others. Some of them, as I have mentioned, still inhabit Central and Eastern Europe, whilst others have a wider distribution on our continent.

This migration must have been an unusually large one. It has been suggested that the Glacial period had some connection with it, and there can be little doubt, as we shall see later on, that a change of climate probably brought about this great Siberian invasion of Europe. But other causes might tend in the same direction, such as want of sufficient food after a few years of great increase of any particular species. It is not known to what we owe the periodic visits of the Central Asiatic Sandgrouse (*Syrrhaptes paradoxus*), Fig. 3, but certain it is that immense flocks of these birds invade Europe from time to time at the present day, just as those mammals may have done in past ages.

The *Siberian* migrations will be spoken of in the subsequent pages, as the Siberian element of the European fauna. These migrations, however, are not the only ones which reached Europe from Asia. The sixth chapter deals with migrations which have

influenced our fauna far more than the Siberian. The
latter did not last long, nor did they affect the whole
of Europe. But what I may call the *Oriental* migrations spread to every corner of Europe and certainly
lasted throughout the whole of the Tertiary Era. The
Oriental element came probably from Central and

FIG. 3.—Central Asiatic Sandgrouse (*Syrrhaptes paradoxus*).

Southern Asia, and in its march to Northern Europe
it was joined by local European migrations. For on
our continent, too, animals originated and spread in
all directions from their centres of dispersal. A
separate chapter has been given to the *Alpine* fauna,
and another to that of South-western Europe, which
will be known by the name of the *Lusitanian* element.

Finally, animals have also reached us from the north, and in the fourth chapter the history of that remarkable migration will be fully discussed under the title of the *Arctic* element of the European fauna.

It is generally believed that Africa played an important rôle in the peopling of our continent, but this is quite a mistake. The eminent Swiss palæontologist Rütimeyer was quite right in saying (p. 42) that it is much more probable that Morocco, Algeria, and Tunis were stocked with animals by way of Gibraltar, and perhaps also by Sicily and Malta, from Europe, than the South of Europe from Africa.

I have already referred to what are known as "centres of dispersion" of animals, but before continuing to explain the general outline of this book, it will be necessary to make a few additional remarks on the subject.

Since every animal naturally tends to spread in every direction from its original home—that is to say, from the place of its origin—the latter should correspond with the centre of its range. And in any particular group of animals the maximum number of species should be formed in the area or zone which is the centre of its distribution. In the great majority of instances this is probably the case, in the higher animals perhaps less so than in the lower; still the rule must hold good that the original home of a species is generally indicated by the centre of its geographical distribution.

Take for example our familiar Badger (*Meles taxus*).

It inhabits Europe and Northern Asia. It is absent apparently from many parts of Central Asia, but it appears again farther south in Palestine, Syria, Persia, Turkestan, and Tibet. West Central Asia would be about the centre of its range. That this corresponds to its place of origin is indicated by the fact that the only three other Badgers known—viz., *M. anakuma*, *M. leucurus*, and *M. albogularis*—are confined to Asia. If we examine the fossil history of the genus, we find that the two most ancient instances of the existence of Badgers have been discovered in Persia, where *M. Polaki* and *M. maraghanus* occur in miocene deposits. The latter had migrated as far west as Greece in miocene times ; no other trace of the Badger, however, is known from Europe until we come to the pleistocene beds. There are a good many cases known among mammals where the centre of dispersion would indicate to us a similar origin. On the other hand, there may be no fossil evidence of the occurrence of a species, or of its ancestors, in Asia, whilst such has been discovered in Europe. I think, however, that the present range of a species forms a safer criterion for the determination of its original home, as the Asiatic continent is still practically unworked from a palæontological point of view. In a letter which I received from Professor Charles Depéret, he advocates the view that the wild Boar (*Sus scrofa*) is probably of European, and not, as I maintained (*c*, p. 455), of Asiatic origin ; because there seemed to be a direct descent from Hyotherium of the middle miocene of

Europe, through the upper miocene Pig of the Mount
Lebéron (*Sus major*) and of Eppelsheim (*Sus
aytiquus*), and the pliocene Pigs of Montpellier (*Sus
provincialis*) and of the Auvergne (*Sus arvernensis*).
No doubt this appears rather a strong case in favour
of the European origin of the wild Boar, but although
the Tertiary strata of Asia, as I remarked, are as yet
little known, a number of fossil pigs are known from
India, Persia, and China, the oldest being the upper
miocene Persian Pig (*Sus maraghanus*). Pigs are
therefore as old in Asia as in Europe, and as a direct
intercourse between the two continents probably never
ceased since miocene times, it is not surprising that
this genus should occur in both. Even if the genus had
its origin in Europe, it is quite possible that in later
Tertiary times, the active centre of origin was shifted
to the neighbouring continent, and that henceforth
many new species issued forth from Asia, some of
which may subsequently have been modified on
reaching our continent. The wild Boar (*Sus scrofa*),
however, to judge from its general range, I must look
upon as merely an immigrant in Europe. I have no
doubt that it originated somewhere in Asia, probably
in the south.

The view I take of the origin of our European
Boar is also supported by Dr. Forsyth Major's recent
researches. He was led to a re-investigation of the
history of the Pig while examining a large number
of fossil skulls in the Museum at Florence, and came
to the conclusion that only three or four species of

recent wild pigs can be clearly distinguished (*b*, p. 298).
One of these, viz., *Sus vittatus*, he thinks, is traceable in slight modifications from Sardinia to New
Guinea and from Japan to South Africa. The centre
of distribution of this species lies in Southern Asia.
Of the three remaining species, two, viz., *Sus ver-
rucosus* and *S. barbarus*, are entirely confined to the
great islands which form part of the Malay Archipelago. Finally, *Sus scrofa*, our Central European
wild Boar, is so closely related to *S. vittatus* that the
Sardinian Boar might be looked upon as a variety of
either the one or the other. At any rate, Dr. Major
recognises clearly in *Sus vittatus* the representative
of the ancestral stock of which *Sus scrofa* is a somewhat modified offshoot.

The fauna of Europe consists, as I have mentioned,
to a large extent of immigrants from the neighbouring
continents. This is especially noticeable among the
higher animals. When we come to the lower, such as
the amphibia, we find a larger percentage, and among
the land mollusca the great majority, to be of European origin. The foreigners are, as we learned, called
Orientals, Siberians, and Arctics. For the sake of
convenience, only two of the great European centres
of origin have a chapter devoted to themselves,
namely, the Alpine and the Lusitanian centres.
There is another, however, of almost equal importance which lies in the east.

In the British Islands there is only an exceedingly
small and insignificant group of species which are

peculiar, and which we may consider to have had their origin there. Almost the whole of the British fauna is composed of streams of migrants which came from the north, south, and east, though many of these immigrant species have since their arrival been more or less distinctly modified into varieties or local races.

The eminent French conchologist Bourguignat (*a*, p. 352) was of opinion that, as far as terrestrial mollusca were concerned, there are in Europe three principal centres of creation or dispersion—all situated in mountainous countries and not in the plains. He distinguished the Spanish, Alpine, and Tauric centres, and believed that almost all species known from Europe had originated in one of these three, and that each of them possessed quite a distinct type of its own. This theory seems to agree very well with the facts of distribution. Let us take, for instance, the genus *Clausilia*, a pretty turret-shaped snail, which abounds on old ruined walls. Only two species, viz., *Cl. laminata* and *Cl. bidentata*, are met with in Ireland. In England we find the same species with the addition of two others, *Cl. biplicata* and *Cl. Rolphii*. Crossing over the Channel to Belgium, these four species occur again, and also several others not known in England. In Germany the list of *Clausiliæ* mounts up to twenty-five species, including all those found in the British Islands. As we proceed eastward the number of species of this genus increases steadily, and when we reach the Caucasus or the Balkan Peninsula the con-

chologist is able to make a collection of several
hundred different kinds, whilst farther east again they
diminish. This clearly indicates there is in South-
Eastern Europe a powerful centre of creation of
Clausiliæ, from which the species have spread all
over Europe. But it is by no means certain that this
centre was always in our continent, for in South-
Eastern Asia and the Malay Archipelago *Clausiliæ*
increase once more. It is interesting to note, however,
that almost all these eastern forms belong to the sub-
genus *Phædusa* (*vide* Boettger), which had only been
known as a fossil genus from a few species in the
Eocene and Oligocene of Southern Europe. The
first centre of origin, therefore, may possibly have
been in Southern Asia, and in these early Tertiary
times a second centre may have become established
in Southern Europe from which the sub-genus *Garnieria* went eastward, *Macroptychia* southward, and
Nenia westward across the Atlantis to South America.
Only a few remnants of these primitive *Clausiliæ*
are now left in Europe, such as the interesting *Cl.*
(*Laminifera*) *Pauli*.

As an example of a genus which has its centre of
distribution in South-Western Europe we might take
that to which our common brown garden slug belongs,
viz., *Arion*. Dr. Simroth, who was the first to point
out that the species of *Arion* had spread over our continent from South-Western Europe (p. 5), is inclined
to the belief that the *Arionidæ* had originated on the
old land-bridge between Europe and North America.

which is generally known by the name of "Atlantis." From this a branch went westward to the New World and another eastward as far as Southern Asia, but *Arion* and a number of other genera are more or less confined to South-Western Europe. Only a few species of *Arion* have a wide range in Europe, one of them, *A. subfuscus*, crossing the borders of our continent into Siberia. In the British Islands and in Western Germany, which are about equi-distant from the supposed creative centre of the genus, there are found five species. In France six or seven species are met with, and in Spain and Portugal about ten. Towards the east, *Arions* diminish in number. This genus, therefore, forms part of a migration which I have designated as "Lusitanian" from *Lusitania*, the name applied by the Romans to what we now call Portugal. Another genus of slugs, *Geomalacus*, is interesting from the fact that one species occurs in the British Islands, being otherwise confined to the Lusitanian province. Parmacella, a slug-like animal bearing a tiny shell at the extremity of its tail, has probably likewise had its origin in this part of Europe. All this, however, will be more fully referred to in the seventh chapter, which deals with the Lusitanian fauna.

As regards the Alpine centre of origin, Dr. Kobelt considers three groups of mollusca as especially characteristic of the Alps, viz., the sub-genus *Campylaea* of the great and widely-spread genus *Helix*, and the genera *Pomatias* and *Zonites*. The latter, which is not to be confounded with our British *Hyalinia* (formerly united with *Zonites*), does not extend very far south or north of the Alps. There may be others too, which owe their origin to these mountains, but most of the terrestrial mollusca are exceedingly ancient, and many genera have existed long before the Alps had made their appearance above the surface of the early Tertiary seas. It should be remembered that *Hyalinia* and *Pupa*, both British genera, are known from carboniferous deposits in forms which closely approach those living at the present day, and in these and a great number of other instances, it is quite impossible to determine the original home of the genus.

This little digression on centres of dispersion will help us to understand in what manner the indigenous element of the European fauna joined in with the alien members as they arrived in our continent. The species confined to South-Eastern England need not necessarily have come to us from Eastern Europe or Siberia. Alpine species spread northward probably at the same time as the Siberian animals went westward. An Alpine form may therefore have joined a batch of the latter and entered England with them. Even a Lusitanian animal may have mingled with these migrants, so that all three elements may occur together in one locality.

But these are exceptions. The migrations have, as a rule, not joined to any great extent; indeed, all those naturalists who have carefully examined the problem of the origin of the European fauna, have felt

that it was composed of elements which arrived at
different times.

The great Russian naturalist, the late Professor
Brandt, distinguished five phases in the history of the
Eurasian mammalian fauna (pp. 249-254). During the
first phase—an uncertain period of long duration—the
mammals held intact their position in the northern
half of Asia. The Mammoth, the Hairy Rhinoceros,
Bison, Musk Ox, Wild Sheep, Reindeer, and perhaps
Tigers, Hyænas, etc., lived then, with numerous
peculiar Rodents, under such climatic conditions,
according to Brandt, that they were able to extend
their range along with tree vegetation to the extreme
north of the Asiatic continent. This, he thinks, seems
to have been the case especially with the Reindeer,
Mammoth, Rhinoceros, and Musk Ox. The second
phase was characterised by the dispersion of the
Northern Asiatic mammalian fauna towards Central,
Southern, and Western Europe, and this period lasted
until the complete extermination of the Mammoth.
The third phase dates from the time when the Mam-
moth and the Hairy Rhinoceros had become extinct,
whilst the fourth commenced with the disappearance
of the Reindeer in Europe, and terminated when the
Wild Ox in the feral state had become unknown.
Finally, the last phase constitutes the present time.
Lartet held similar views, and also believed that
Europe was peopled by successive migrations from
Asia.

Botanists have worked at the problem of the

European flora much more systematically, and our knowledge of the origin of that flora has been greatly increased within the last twenty years, chiefly ,by the researches of Professor Engler. More recently, detailed studies have been made in Scandinavia by Professor Blytt, in the Alps by Dr. Christ and Mr. Ball, in Germany by Professor Drude, Dr. Schulz, and many others. Dr. Schulz (p. 1) is of opinion that the great majority of the European plants have either migrated to or have originated in our continent since the beginning of the Pliocene epoch, and that the original home of the immigrants must be looked for in Asia and in Arctic America. From the latter an almost uninterrupted migration must have taken place during the greater part of Tertiary times up to the commencement of the Pliocene epoch, partly over a direct land-connection with Europe by way of Greenland, Iceland, and the Faroes, and also *via* Spitsbergen, Franz Josef Land and Novaya Zemlya, and partly by an indirect one across the Behring Straits between Alaska and Kamtchatka.

A good deal of work still requires to be done before zoologists have acquired the same intimacy with the European fauna as botanists have with the flora. However, the view that our animals all come from Asia, as was long ago believed, has been abandoned for some time. The first to bring under the notice of naturalists the hypothesis, that there must have been two distinct migrations of northern animals to Central Europe—one from the north, and another from the east—was the late Mr. Bogdanov. The Arctic species, of which remains have been discovered in the Pyrenees—namely, the Reindeer, Arctic Hare, Willow Grouse, etc., he thought had nothing to do with those which invaded Europe from Siberia during the Glacial period. He maintained that the former had quite a distinct origin, and came from Scandinavia (p. 26).

As I shall deal with this problem more fully in a subsequent chapter, I need only mention that I fully agree with the view expressed by Mr. Bogdanov that two distinct migrations of northern species to Central Europe can be traced.

No one, I think, has done more in fostering a careful study of the migrations of animals than our distinguished geologist Professor Boyd Dawkins. He did not follow Bogdanov in distinguishing two Arctic migrations; however, he did more in constructing a very ingenious chart (*a*, p. 111) representing the geography of Europe during the last and most recent geological epoch—the Pleistocene—and indicating on it the probable extent, during that time, of an eastern and a southern migration of mammals. The map is very instructive, and is the first ever published giving a clear idea of a southern and an eastern migration to Europe. He believed that the migration of the southern mammals northward, took place conjunctly with the westward movement of the eastern species. Having once reached Europe, the southern species are supposed to have passed northward in summer time, whilst the eastern forms (he calls them northern) would swing southwards. The two migrations would thus occupy, at different times of the year, the same tract of ground (*a*, p. 113). From the mingling of the remains of the Hyæna with those of the Reindeer and Hippopotamus in the Kirkdale Cavern, he infers that the former preyed upon the Reindeer at one time of the year, and on the Hippopotamus at another. He argues that in such a manner might be explained the curious mixture of northern and southern types which we find in the British pleistocene and in cave deposits.

Besides mammals, the only European animals which have received some attention with a view to a study of their origin, are the Butterflies and the Land-Snails. The entomologists who have taken up the problem have in so far scarcely produced satisfactory results, as they all seemed to be bound down to the hypothesis that practically all the butterflies had been destroyed in Europe during the Glacial period. Hofman, in his interesting little work, comes to the conclusion (p. 50), that only in Greece and Spain could a small remnant of the butterflies have survived the extreme rigours of climate. Greece was at that time connected with Asia Minor, and Spain with North Africa; and the author supposes that the semi-alien fauna inhabiting these tracts was mainly responsible for the re-stocking of Southern Europe, but that the main mass of our butterflies are post-glacial Siberian immigrants.

The work published by Messrs. Speyer deals only
with the origin of the Central European Butterflies.
The period during which our European species
originated is not specified, but the authors believe
that they had their home either in Southern Russia
or Central Asia. The fact that the number of
butterflies decreases very considerably as we proceed north-westward in Europe appears to them
to substantiate these views. The apparent dislike
evinced by butterflies to the damp Atlantic Coast
climate, they think, clearly indicates that they had
originated in a dry and more continental climate.
The history of the North European Butterflies
and Moths has been carefully described by Mr.
Petersen. He adopts Hofman's theory as to the
almost total extinction of the Lepidoptera in
Europe during the Glacial period. The chief immigration to Europe after that period is, he thinks,
Siberian.

At first there appeared species which belonged
to a cold climate, and which now live in elevated regions; then came forms suited to a milder
climate, which established themselves on the north-
easterly slopes of the Alps. The most recent addition
which our continent has received from Siberia is,
according to Mr. Petersen, the present Scandinavian
fauna. Scandinavia has obtained a larger number of
species than the European plain, because to this last
migration were added such as prefer a northern or
Alpine climate.

As a contribution to the history and composition of
the European fauna, by far the most important work
ever published is that of Dr. Kobelt, the eminent
German conchologist. Whilst the researches into the
origin of the Lepidoptera, above described, have been
marred by the prevalent prejudice as to the dele-
terious effects of a glacial climate on the butterflies,
the present author boldly works out the problem on
independent lines. He shuns theories and speculations almost altogether. His great work, as yet
practically unknown, the result of a lifetime of the
most painstaking labour, ranks among the most important contributions to zoogeography. I shall have
frequent occasion to refer to it throughout these pages.
Meanwhile some of his more remarkable conclusions
may be mentioned. "Comparing all classes of
animals as to their zoogeographical importance, the
highest rank must undoubtedly be accorded to the
land-snails" (i., p. 7). "The Pleistocene, and with it
the land and fresh-water molluscan fauna of the
present day has been gradually evolved from the
Tertiary one, and its roots can be traced through the
Cretaceous to the Jurassic epoch. During the whole
of that time no sudden appearance of a new fauna
can be demonstrated. Quite slowly, step by step, the
Cretaceous is succeeded by the Tertiary fauna, and
one after the other of the characteristic palæarctic
genera appear—first the fresh-water, then the land
forms" (p. 141). "The division of the North Alpine
from the South Alpine fauna must be older than the

Glacial period ; and the present Central European fauna had already become developed from the Pliocene *in all its details of form and distribution* before the commencement of the Ice Age" (p. 162). "We must draw the conclusion from the preceding remarks, that the present (palæarctic) molluscan fauna in its distribution is older than the Glacial period, and that the latter produced merely a retreat of the fauna from the most inhospitable regions of Europe with a subsequent re-immigration, but did not cause its destruction" (i., p. 169).

A few attempts have also been made by naturalists to trace the origin of the fauna of some smaller European areas. Thus Rütimeyer, in dealing with the mammalian fauna of Switzerland, remarks (p. 31) "that it seems certain that, in spite of many local disturbances, the continuity of generations was never interrupted throughout the whole of the Tertiary period until the present day."

An even more interesting memoir is that of Mr. Köppen on the origin of the Crimean fauna. It is only recently, according to this author, that this peninsula has become connected with Southern Russia. And it is for this reason that the Squirrel and a number of other animals, and also plants, present in Russia, are absent from the Crimea. Originally the latter probably formed a westward continuation of the Caucasus, and at that time it was surrounded by the sea on all other sides. "Much later," he continues, "after and in consequence of a local subsidence, the country between the Caucasus and the Crimea became interrupted. The latter existed for a long time as an island, and only much later, in recent geological times, did it become united with Southern Russia by means of the isthmus of Perekop."

There is, on the whole, a great diversity of opinion as to how the European fauna has originated ; however, except in Dr. Kobelt's work, no attempt has hitherto been made to collect together all the available information, and to include in the inquiry more than one class of animals. The little work which I venture to bring before the public will not by any means exhaust the subject, nor is our knowledge of the European fauna sufficient to give more than a mere sketch of many of the animal groups mentioned. As we have learned in the introduction, different classes of animals are not all of equal importance in indicating the changes which have taken place in the distribution of land and water. While Dr. Kobelt is of opinion that the land-snails are by far the most important in such an inquiry, Mr. Lydekker believes that mammals afford the safest and truest indications of such changes. Mr. Beddard puts in a claim for earthworms, as even a narrow strait of sea-water forms an insuperable barrier to their dispersion. Dr. Wallace agrees with Mr. Lydekker, and goes so far as to say (p. 74) that "whenever we find that a considerable number of the mammals of two countries exhibit distinct marks of relationship, we may be sure that an actual land-connection, or at

all events an approach to within a very few miles
of each other, has at one time existed." Besides
the groups referred to, I claim that particular
attention should be devoted to Amphibia, which,
contrary to Wallace, I hold do not possess special
facilities for dispersal; and also to spiders and to
all wingless animals leading a subterranean life, such
as some of the wood-lice, planarian worms and
apterous beetles.

A thorough knowledge of the changes in the distribution of land and water is desirable in order to
appreciate the extent and variations of former migrations. A study of the British fauna, for example,
teaches us that the British Islands were once connected with one another and with the continent of
Europe between England and France. It was
Professor James Geikie, I believe, who first pointed
out, many years ago, that the area now covered by
the Irish Sea was formerly in all probability a freshwater lake. This had its outlet at the southern extremity in the form of a stream into which most likely
flowed the smaller rivers from the south-east of
Ireland, and which was joined from the east by the
Severn, and finally debouched into the Atlantic
(Fig. 4). The range in the British Islands of those
species which have migrated to them from the south,
indicates that whilst the Atlantic Ocean had gradually
crept up and flooded the area between Ireland and
Wales, and had turned the fresh-water lake into a bay,
communication between Scotland and Ireland was

still possible. The occurrence of many Scandinavian
species in Scotland which are absent on the continent

FIG. 4.—Map of the British Islands and surrounding area at a time
when the earlier members of the southern migration reached
England. (Only some of the rivers have been indicated. The
shaded parts represent water, the light land.)

of Europe, indicates that these two countries also
were united formerly. Most geologists hold that such

a connection, if it existed, must have broken down in Pliocene times. Professor Judd, however, has ex-pressed his belief (p. 1008) that it still existed until after the appearance of man in Northern Europe, and that our forefathers might have been able to walk dry foot from Scotland to Norway.

I shall also show on distributional evidence, in the fourth chapter, that until recent geological times Scandinavia was continued northward, by way of Bear Island, with Spitsbergen and probably Franz Josef Land, which islands again were joined with North Greenland and Arctic North America, and that the polar fauna and flora were able to spread on this land-connection to both America and Europe.

That Gibraltar was connected with Morocco, and Sicily with Southern Italy and Greece on the one hand, and with Tunis on the other, is more generally recognised; whilst Professor Suess has shown (vol. i., p. 442), on purely geological grounds, that the Egean Sea was dry land up till quite recently— certainly, he thinks, till after the appearance of man. This supposition enables us to understand, as will be more fully discussed in the sixth chapter, how the Oriental fauna entered Europe. Such minor zoo-geographical problems as the occurrence of the Wild Goat of Asia Minor (*Capra ægagrus*) on the islands of Crete and on some of the Cyclades now almost explain themselves. The Sea of Marmora is prob-ably a modern formation, so that Asia Minor ex-tended not long ago beyond the Turkish capital, but

Dr. Kobelt believes that an arm of the Black Sea communicated up till recent times along the lower course of the Maritza with the Gulf of Saros. It can be shown also that Sardinia and Corsica formed part of the continent of Europe, and that their present fauna and flora reached them by migration on land.

The Russian naturalists, Brandt and Köppen, believed that at no very distant date a sea extended right across Eastern Russia from the Caspian to the Arctic Ocean, whilst Professor Boyd Dawkins expressed himself in very similar language as follows (6, p. 35): "Before the lowering of the temperature in Central Europe the sea had already rolled through the low country of Russia, from the Caspian to the White Sea and the Baltic, and formed a barrier to western migration to the Arctic mammals of Asia." These naturalists based their opinions on distributional evidence, but additional facts will be brought forward in the fifth chapter to substantiate these views.

These are some of the more important geographical events which will be dealt with in detail in the subsequent chapters in connection with the history of the migrations of the European fauna.

A separate chapter has been devoted to the British fauna and its origin, since it plays a very important part in the evolution of that of our continent. So essential is a thorough knowledge of this fauna, that I think it would be difficult to understand, without

it, the main features of the great migrations; and I
have before now expressed the opinion that the
British fauna forms the key to the solution of the
problem of the origin of European animals. We
know that our British species came to us by land
—at least the bulk of them. But we want to
know what direction they came from, and at what
time they arrived. When Ireland became disconnected from Great Britain, and the latter from
Scandinavia and France, is another interesting
problem. Professor Boyd Dawkins has indicated
to us a method of the special line of research to
meet such inquiries. "The absence," he says (b,
p. xxix), "of the beaver and the dormouse from Ireland must be due to the existence of some barrier
to their westward migration from the adjacent mainland, and the fact that the Alpine hare is indigenous, while the common hare is absent, implies
that, so far as relates to the former animal, the barrier
did not exist."

Many members of the great Siberian invasion
reached England, but Ireland remained entirely
free from these migrants. The assumption therefore seems not unreasonable, that the latter country
at the time of their arrival was no longer joined
to England. The great bulk of the Irish fauna
is composed of Lusitanian, Alpine, and Oriental
immigrants, and there is besides a distinctly Arctic or
North American element. All these, of course, must
have established themselves in Ireland before the

Siberian fauna set foot in England, since it has been shown that a continuous land-surface was necessary for their migration. Owing to the perfect preservation of the remains of the Siberian migrants in recent continental deposits, the history of that migration can be clearly followed, and it is possible even to determine the date of its arrival in England—in geological language at any rate. The time of the colonisation of Ireland can be thus approximately fixed as having taken place at a period prior to the arrival of the Siberian migrants in England.

All those who have seriously studied the problems presented by our British fauna—notably the late Professor Forbes, and more recently Mr. Carpenter and myself—are agreed that the Lusitanian element is the oldest, and that the newest is that which has come to us from the east.

The sequence of events in the British Islands was probably as follows :—The first comers were the members of that fauna which issued from South-western Europe; then came the Alpine, and at the same time probably the Arctic and the Oriental; and finally the Eastern or Siberian. The migrations of all but the last continued, uninterruptedly, for very long periods.

The study of these migrations has convinced me that, though climate was a powerful factor in the evolution or history of the European fauna, the geographical changes which took place on our continent in later Tertiary times exerted a yet stronger influence. The principal climatic disturbance is generally supposed to have been the so-called "Ice Age." So firmly rooted is the conviction, among naturalists of the present day, of the enormous destruction which this period produced on our European fauna, so that all animal life practically disappeared from large areas of our continent, that it is desirable that we should now shortly review the history of that remarkable period in order to ascertain in how far these views are corroborated by facts. Frequent reference, moreover, will be made throughout this work to the theories connected with the Glacial period.

It has been stated by an eminent geologist that during part of the Glacial period the climate was such that neither plants nor animals could have existed in the British Islands. If that had been so, it is evident that very few organisms could have even survived in France, though a number of Arctic species might have dragged on an existence in Southern Europe. At any rate, on the return of more genial conditions, the Arctic species would undoubtedly have been the first to gain admission to the British Islands, to re-people the arid wastes. Our supposition that the Lusitanian element in the British fauna is the oldest would therefore be wrong. From early Tertiary times onward, the climate of Europe, which was then semi-tropical, gradually became more and more temperate; until finally the Ice Age or Glacial period arrived, during which, according to Professor

J. Geikie—one of our highest authorities on this subject—a great part of Northern Europe became practically uninhabitable owing to the severity of the climate.

To enable us to judge better of the true value of the many hypotheses which have been advanced to account for this supposed extraordinary fall of temperature during the "Ice Age," we must compare the views of other authorities with the one just quoted. I do not propose to discuss the causes which have led to the production of the Glacial period— those interested in these questions should consult the writings of Dr. Croll, Professor J. Geikie, Professor Bonney, Mr. Falsan, and others—but merely to give the climatic aspects from a physical, zoological, and botanical point of view.

According to Professor Penck (*a*, p. 12), the nature of the glacial climate can be determined by comparing the snow-line of the Glacial period with that of the present day. The position of the snow-line is dependent on two climatic factors—viz., precipitation and temperature. We know the height at which snow must have lain permanently during the Glacial period, or during the maximum phase of glaciation. If the Ice Age had been produced solely by an increase of snowfall, as has been suggested, Professor Penck tells us that then it must have snowed three or four times as much as it does now. But he does not adopt the view that the Ice Age is due to an increase of snowfall alone. His calculations, based upon the

height of the snow-line, tend to show that a general
decrease of temperature to the extent of from 4-5
degrees Centigrade (all other atmospheric conditions
remaining the same as now) would be sufficient to
give us back the Glacial period.

Professor Neumayr (p. 619) adopted a similar principle in determining the temperature which prevailed
in Europe during the Glacial period. Snow now lies
in the Pyrenees 1000 metres higher than it did then,
1,200 metres higher in the Alps, and 800 metres
higher in the Tatra mountains. Since the temperature in Central Europe decreases by half a degree
Centigrade for every 100 metres of elevation, it follows
that if the glacial phenomena had only been brought
about by a decrease of temperature without an increase of moisture, we should have had a reduction
of temperature during the Glacial period of six
degrees Centigrade in the Pyrenees, of seven degrees
in the Alps, and of four in the Tatra mountains.
The general lowering of the temperature of Europe,
says Professor Neumayr, could not have amounted to
more than six degrees Centigrade. Moreover, he is
of opinion that the very low snow-line in the British
Islands proves that even during the Ice Age a comparatively mild climate prevailed there, and that
the climatic conditions generally, in the different
parts of Europe, were relatively about the same as
they are now.

Professor J. Geikie does not give us his views as to
the temperature of the Glacial period, but he main-

tains that a lowering of the temperature is evinced
not only by the widespread phenomena of glaciation,
but by the former presence in our temperate latitudes
of a northern fauna and flora.

Mr. Charles Martins, who based his calculations
on the temperature during the Glacial period on
the glaciers of Chamounix, concluded that it only
needed a lowering of the temperature to the extent
of four degrees Centigrade to bring the glaciers down
to the plain of Geneva, and in fact give us back the
Glacial period. It need not surprise us, therefore,
that the French geologist, Mr. Falsan, the author
of *La période glacière*, is of opinion (p. 230) that the
mean annual temperature of France during the
Glacial period was approximately from 6-9 degrees
Centigrade, perhaps more. Close to the immense
glaciers of the Rhone, it might have been about
six degrees. This is the actual mean annual temperature of the South-west of Sweden and Norway, or
the North of Scotland.

Although all these investigations tend to show
that the climate of Europe during the Glacial period
was by no means so severe as we are often led to
believe, yet there exists also a school of geologists
who maintain there was actually a higher temperature
than at present. The inconsistency of mentioning
heat in connection with ice and snow is more
apparent, however, than real, for we must remember
Tyndall's original remark on this subject. It is the
snow, he says, which feeds the glaciers. But the

snow comes from the clouds, and these again originate from the vapours which the sun causes to be absorbed from the ocean. Without the sun's heat, we should have no water vapour in the atmosphere; without vapour, no clouds; without clouds, no snow; without snow, no glaciers. The ice of glaciers, therefore, owes its origin indirectly to the sun's heat. It has been supposed that if the sun's heat diminished, larger glaciers would form than those existing to-day, but the diminution of the solar heat would infallibly reduce the amount of water vapour in the air, and would thus stop the very source of glaciers.

Mr. Falsan even admits that without a change of the mean annual temperature (p. 201) of Europe, the central portions of our continent might at this period have enjoyed an insular climate. This more equable and humid climate could, within certain limits, favour the development of the ancient glaciers by increasing the snowfall and slackening the summer rate of melting.

It seems evident then, according to these views, that with a comparatively slight change of the atmospheric conditions in the British Islands, we might have glaciers back again on all our highest mountain ranges in England, Scotland, and Ireland. But a widespread belief seems to prevail that the presence of glaciers implies a very low temperature. Snow and ice, however, are formed as soon as the temperature falls below freezing point; it does not matter whether there be 1 or 20 degrees of cold. Winters with a few degrees of frost will be just as favourable for the growth of glaciers as winters with the most severe cold.

Let us now see what the fauna and flora, as far as we know it, tell us of the climate of the Glacial period. At the very outset of our inquiry we are confronted with one very serious difficulty in the problem, and that is the supposed occurrence of inter-glacial mild phases alternating with colder ones during the Ice Age. At first, when traces of a temperate flora and fauna were discovered intercalated between two layers of boulder clay, their presence was explained by the supposition of a mild inter-glacial period. The famous Forest-bed on the east coast of England was also pronounced to be an inter-glacial deposit, though not coming precisely under this definition. In a few places one such bed was found, in some two or more, and in others none at all. Professor James Geikie discovered the evidences of no less than five of such inter-glacial epochs (p. 612) in Europe. Lest a reader of that author's remarkable work on the Ice Age might carry with him the idea that his hypotheses had met with general acceptance, a few quotations from almost equally high authorities on glacial matters will be useful. "That the glaciers," remarks Professor Bonney (p. 245), "were liable to important oscillations seems to be proved, but whether the evidence suffices to establish inter-glacial epochs, in the usual sense of the words, is more doubtful.

When the snow-fields, as in the Alps, were much more extensive than they are at present, the glaciers which radiated from them would be more sensitive to 'minor climatal change. Even now they oscillate considerably. But during a Glacial epoch, an inch, either more or less, of precipitation might mean a considerable advance or retreat of the ice in the lowlands." French geologists look with even less favour on Professor Geikie's theories. Mr. Falsan (p. 212) says that he agrees with Messrs. Favre, de Saporta, Lory, de Mortillet, Desor, de Lapparent, Lortet, Chantre, Benoit, Fontannes, Depéret, and many other geologists, that there was only a single Glacial period, which, according to each particular region, might be divided into several phases, or into their equivalents—viz., one or more extensions of the ancient glaciers. But, on the whole, the view that there was at least one inter-glacial phase in the Glacial period meets with more general acceptance among geologists, I think, though the other opinion agrees much better with the nature of the fauna and flora as it has been revealed to us from the pleistocene deposits.

The occurrence of the remains of such arctic species of mammals as the Musk-Ox, Arctic Fox, Glutton, Lemming, and many others in these deposits, is frequently held up to us by geologists as a proof of the prevalence of an arctic climate while these beds were laid down. And indeed this appears at first a most satisfactory explanation of the

phenomenon. But we must not judge the climate
of Europe by their presence alone. As I shall
explain more fully in Chapter V., these species
invaded Europe owing to two circumstances. Firstly,
because the climate of Siberia was becoming colder,
necessitating a southward movement, with a consequent over-population in a reduced area; secondly,
because a new short route to Europe had been
opened up for them about the same time (see
p. 221). An invasion of Europe therefore took place
from east to west. Similar invasions occur even at
the present day, though not caused by a change in
our climate, for every now and then immense flocks
of the Siberian Sandgrouse emigrate to our continent.
The mammalian migrants referred to are not to be
looked upon as constituting the whole of our fauna
at that time. Europe had a fauna of its own, and
these invaders merely mingled with our animals.
There was, no doubt, a keen struggle for existence,
as the result of which the weaker in many cases
succumbed. The hypothesis, however, that these
Siberian migrants occupied an empty continent,
forsaken by its pre-glacial inhabitants, is not supported by any facts.

All those who have investigated the pleistocene
fauna have been struck by the extraordinary mixture
of northern and southern types of animals. Professor
Dawkins attempted to explain these facts by the supposition (p. 113) that "in the summer time the southern
species would pass northwards, and in the winter

time the northern would sway southwards, and thus occupy at different times of the year the same tract of ground, as is now the case with the elks and reindeer." "In some of the caverns," he continues (p. 114), "such as that of Kirkdale, the hyæna preyed upon the reindeer at one time of the year, and the hippopotamus at another."

A similar mingling of northern and southern faunas has also been observed in France. Mr. Falsan tells us (p. 236), that the remains of the mammals gathered and determined by Lartet and Gaudry belong partly to species which have been wrongly regarded as indications of a severe climate, and partly to such as are accustomed to a relatively mild temperature. In several localities in France, viz., at Levallois, St. Acheul, and Arcy, the remains of the Hippopotamus have occurred together with those of the Reindeer; whilst, according to Sir H. Howorth, the Lion has been found together with northern Voles at Bicêtre, near Paris. It is stated by the same authority (p. 115) that much the same conditions exist in Germany. "The lion and the spotted hyæna, the mammoth and rhinoceros, were found with the marmot, the suslik, the lemming, the pica, and the reindeer." At another locality near Thiede, remains of the Mammoth, woolly Rhinoceros, Horse, Ox, Reindeer, Arctic Fox, Lemming, and Pica are met with in the same deposit. In quoting the presence of these northern animals in Europe as evidence of an arctic climate, we commit a fatal mistake. Indeed,

breeders of animals and those acquainted with zoological gardens know perfectly well that it is much
easier to keep a northern species in a southern
climate, than a southern species in a northern one.
If in a Central European deposit occur a mixture of
northern and southern forms of animals, the presence
of the latter is more remarkable than that of the
former. Logically, we should look upon the occurrence
of southern species in the north, therefore, as support-
ing the view that a mild climate had induced them
to travel northward. The only indication, indeed,
of the presence of a Monkey in the British Isles
in former times comes to us from the very same
strata which have also yielded the remains of the
Siberian mammals.

Before I conclude the consideration of the pleistocene fauna, it may be of interest to hear what
Mr. Lydekker, one of our highest authorities
on fossil mammals, has to say on this subject.
"The most remarkable feature connected with
this fauna is the apparently contradictory evidence
which it affords as to the nature of the climate
then prevalent. The Glutton, Reindeer, Arctic
Fox, and Musk-Ox are strongly indicative of
a more or less arctic climate; many of the Voles
(*Microtus*), Picas (*Lagomys*), and Susliks (*Spermophilus*), together with the Saiga Antelope,
appear to point equally strongly to the prevalence
of a Steppe-like condition; while the Hippopotamus
and Spotted Hyæna seem as much in favour of a

sub-tropical state of things. Many attempts have
been made to reconcile these apparently contradictory circumstances; one of the older views being
that while the tropical types of animals lived during
a warm interlude, they migrated southwards with
the incoming of colder conditions to the arctic type
of fauna. Since, however, it has now been ascertained
that the remains of both tropical and arctic forms
have been found lying side by side in the same bed,
it is perfectly certain that such an explanation will
not meet the exigencies of the case" (p. 300).

In Germany the remains of the Siberian mammals
occur to a large extent in a pleistocene deposit known
as "loess," and the theory has of late years gained
ground that the latter is the fine dust-like sand
accumulated during an intensely arctic dry climate.
That many of the mammals discovered in the "loess"
now inhabit the dry steppes of Eastern Europe and
North-Western Asia seems to lend support to this
supposition; but besides the mammals there are also
land and freshwater shells in this deposit. The molluscan fauna certainly indicates no steppe-character,
according to Dr. Kobelt (*b*, i. p. 166).

The attempt to utilise the Siberian migrants to
Europe as indicators of a severe climate there, certainly fails to establish conviction. But it may be
asked, surely the remains of the Alpine and Arctic
plants which have been found in pleistocene deposits
must decide this question in favour of one or the
other hypothesis? Let us test it.

Plants being more directly affected than animals by changes of temperature and rainfall, remarks Mr. Clement Reid (p. 185), give evidence of the highest value when we inquire into former climatic conditions. The severity of the climate during the Glacial period is often assumed from the occurrence in pleistocene strata of such plants as *Dryas octopetala*, some species of willow, the dwarf birch, and others, which are now found in high latitudes and in the Alps, but are, as a rule, absent from the plain of Northern Europe. Professor J. Geikie goes so far as to state (p. 398) that it was unlikely that southern England during the climax of the glacial cold had much if any vegetation to boast of, and continues, " It is certain, however, that it was clothed and peopled by an Arctic flora and fauna when the climatic conditions were somewhat less severe, relics of that flora having been detected at Bovey Tracey." He believes, therefore, that an Arctic flora took possession of England as soon as the climate enabled it to live in the country. Arctic plants, according to this explanation of the sequence of events, were the first immigrants to reconquer the dreary, plantless wastes and make them habitable for mammals.

Fortunately these views do not at all agree with those of many of our leading European botanists and others entitled to have a voice in the matter. Professor Warming is of opinion that the main mass of the present flora of Greenland survived the Glacial period in that country (p. 403); whilst Professor

Drude has shown (p. 288) that all plant life could not possibly have been destroyed in northern countries. He maintains that the greater part of the Arctic floral elements which unite Greenland and Scandinavia must have survived the Glacial period in these countries in sheltered localities. Indeed, he justly remarks, where at the present moment do we find such plantless wastes? Greenland, Franz-Josef Land, and Grinnell Land, situated in high Arctic latitudes, all have a flora composed of flowering plants and cryptograms. "I cannot understand," he continues (p. 286), "why a flora, possibly mixed with northern forms but in the main points agreeing with our present floral elements, should not have persisted throughout the Ice Age even in the heart of Germany." "To my mind," says Col. Feilden, the well-known Arctic traveller (b, p. 51), "it seems indisputable that several plants now confined to the polar area must have originated there, and have outlived the period of greatest ice-development in that region." The theory in favour of a survival of the pre-glacial flora has been especially strengthened by the late Mr. Ball (than whom probably no botanist possessed a better knowledge of Alpine plants), who was strongly in favour of this view as far as the Alps are concerned. "Is it credible," he says (p. 576), "that in the short interval since the close of the Glacial period hundreds of very distinct species and several genera have been developed on the Alps, and, what is no less hard to conceive, that several of these non-Arctic species and genera should still more recently have been distributed at wide intervals throughout a discontinuous mountain chain some 1,500 miles in length, from the Pyrenees to the Eastern Carpathians?" Mr. Ball's remarks, indeed, just touch upon a very important characteristic of all the so-called *Alpine plants*. In Europe they chiefly occur in Scandinavia and the central and southern mountain ranges, whilst they are mostly absent from the intervening lowlands. Again, we find a large number of species in the mountains of Central Asia and in some of the North American mountains. Almost all species of Alpine plants, in fact, are examples of discontinuous distribution; and this, as every naturalist knows, is always, in both animals and plants, a proof of antiquity.

The glacial or Alpine flora is very old, and must have originated long before the Ice Age. But it might be urged, why should these plants be now almost confined to the Arctic regions and the higher mountain ranges, where the temperature undoubtedly is very low, if they had originated during a pre-glacial period probably much milder than the present? The answer can be given by those who have made Alpine plants their special study, and who have attempted to grow them by administering to them a temperature and such climatic conditions as to be most conducive to good health. We should all expect these plants to be very robust, and especially to be able to stand extremely low temperatures. But, strange to say, the very opposite is the case. Professor Blytt tells us (p. 19) that "Arctic and Alpine species in the Christiania Botanic Gardens endure the strongest summer heat without injury, while they are often destroyed when not sufficiently covered during winter." The English climate then, one would think, ought to suit these plants, since the winters are not too cold; but we find that at Kew Gardens the large collection of Alpine plants have to be wintered in frames under glass in order to keep them in good health; and Professor Dyer, the Director of the Gardens, thinks they are mostly intolerant of very low temperatures (compare also pp. 161-164).

Such being the constitution of Alpine plants, how could they possibly have originated during the Glacial period and wandered from the mountains into the plains, across numbers of formidable barriers, often exposed to icy winds, for thousands of miles? As a matter of fact, Alpine plants have survived in the high North and in the Alps because they are there permanently protected during winter by a covering of snow from very low temperatures, and they are at the same time prevented from drying up. If they are given sufficient moisture and a constant, mild temperature they seem to do very well. Such conditions are afforded them in many parts of the British Islands, and we find indeed the Mountain Avens (*Dryas octopetala*), one of the most typically Arctic plants, growing wild in profusion on the coast of Galway, in Ireland, at sea-level. The winter temperature of that part of Ireland resembles that of southern Europe, being no less than 12° Fahr. above freezing point. This fact appears to strengthen the view not only that the Alpine flora is of pre-glacial origin, but that the climate of Europe during the Glacial period was mild.

Having now shortly reviewed the state of our knowledge with regard to the former presence in our temperate latitudes of Arctic animals and plants, it still remains for me to give a succinct statement of the light thrown by this fauna and flora on the wide-spread phenomena of glaciation. It is necessary to do so, because, though the greater development of glaciers on the mountains of Europe in former times does not presuppose the prevalence of an Arctic climate, the survival through the Ice Age of a fauna and flora could not possibly have taken place in northern Europe if the theories of glaciation now so much in vogue are really true. Professor Geikie reminds us, in speaking of his native country (p. 67), that "we must believe that all the hills and valleys were once swathed in snow and ice; that the whole of Scotland was at some distant date buried underneath one immense *mer de glace*, through which peered only the higher mountain tops." That under such conditions no fauna or flora to speak of could have survived in Scotland is evident. Then again he argues (p. 426) that because in the great plain of Europe we meet occasionally with striated rock-surfaces and *roches moutonnées* very similar. to those produced by the glaciers of Switzerland, it must have been traversed by "inland ice" flowing from Scandinavia and the Baltic southward. The boulder clay of Germany is supposed to have accumulated underneath this vast "*mer de glace*," as he calls it. There is no question here of a simple local development of glaciers, such as could have existed under a mild and moist climate; practically all the plants and animals would have been annihilated in northern Europe under such conditions, as there were no areas free from ice. A more vivid idea of the state of Europe during the epoch of maximum glaciation will be obtained by looking at Professor Geikie's map (p. 437). The whole of Scandinavia, Iceland, Scotland, Ireland, and Switzerland is there represented as having been completely enveloped in ice, and also the greater part of Russia, Germany, and England. In speaking of Scandinavia (p. 424) he remarks that "the whole country has been moulded and rubbed and polished by one immense sheet of ice, which in its deeper portions could hardly have been less than 5000 feet or even 6000 feet thick." The greater portion of the area indicated as having been underneath a sheet of ice is thickly covered with superficial accumulations of gravel, sand, and clay. The latter is generally spoken of as "boulder clay," and, with the associated sand and gravel, it may be observed equally well in Russia or Germany, in England or Ireland. As a rule these stony clays thicken out as they are traced from the high-lying tracts to the low grounds; and especially near the mountains the rock-surfaces are often polished and striated. "For many years it was believed," continues Professor Geikie (p. 432), "that all those superficial deposits were of iceberg origin. The low grounds of Northern Europe were supposed to have been submerged at a time when numerous icebergs, detached from glaciers in Scandinavia and Finland, sailed across the drowned countries, dropping rock-rubbish on the way. Such was thought to have been the origin of the erratics, stony clay, and other superficial accumulations, and hence they came to be known as the 'great northern drift formation.'" "But," he adds (p. 433), "when the phenomena came to be studied in greater detail and over a wider area, this explanation did not prove satisfactory. The facts described in the preceding paragraphs—the occurrence of striated surfaces and *roches moutonnées*, the disturbed appearances associated with the till, and the not infrequent presence of giants' kettles —convinced geologists that all the vast regions over which boulder-clay is distributed were formerly occupied by the 'inland ice' of Scandinavia."

I think Professor Geikie over-estimates the value of the evidences which appear to be in favour of his theory. His treatise on the Ice Age leaves one under the impression that the older view of the marine origin of the boulder-clay is not only

done with for good and all, but that no geologists
nowadays believe in it. If a more careful study of
the glacial phenomena has led most geologists to
abandon what I might call the "marine view" in
favour of the terrestrial one, a more careful study
of the fauna and flora will, I venture to think, have
the opposite effect. However, it appears that even
from a purely geological point of view more can
be said in favour of the old theory than Professor Geikie and his school are ready to admit.
Thus we are told by Professor Bonney (p. 280),
in referring to the boulder-clay, that "the singular
mixture and apparent crossing of the paths of
boulders are less difficult to explain on the hypothesis of distribution by floating ice than on that of
transport by land-ice, because, in the former case,
though the drift of winds and currents would be
generally in one direction, both might be varied at
particular seasons. So far as concerns the distribution and thickness of the glacial deposits, there is not
much to choose between either hypothesis; but on
that of land-ice it is extremely difficult to explain
the intercalation of perfectly stratified sands and
gravel and of boulder-clay, as well as the not infrequent signs of bedding in the latter." "Anything,"
writes Professor Cole (p. 239), "that keeps open the
position maintained by Lyell and others, that
extensive glaciation is compatible with mild and
sheltered nooks and corners, and that much of
the distribution of boulder-clay was performed in

seas and not on land, may be welcomed by
rationalists, at any rate until further research has
been carried on among the Arctic glaciers. At
present every year brings evidence of modern
marine boulder-clays in high latitudes, and removes
us farther and farther from belief in a *moraine
profonde.*" That foraminifera are occasionally found
in boulder-clay has been known for a long time,
but it is only within recent years that these marine
organisms have been shown to occur in so many
localities, that Mr. Wright, who examined a large
number of samples, says (p. 269), "I am forced
to the conclusion that the Scottish as well as
the Irish boulder-clay is a true marine sedimentary
deposit"

In the fourth and fifth chapters I shall return to
this subject again, and mention a number of facts
of distribution which appear to me much easier of
explanation by means of the marine than by the land-
ice theory. But I do not propose to go into further
geological details in this volume, as I think I have
clearly conveyed my position in this controversy.

Before concluding this short review of the glacial
problem, so far as it affects the origin of the European
fauna, I should like to refer to the opinion of one
who has devoted years to the study of the glacial
phenomena in the Arctic Regions, viz., Col.
Feilden. "To a certain extent," he says (*a*, p. 57),
"all boulder clays at home are fragmentary when
compared with the boulder-bearing beds of Kolguev,

which we may safely assume are 50 miles in length
by 40 in width, with a thickness of not less than 250
feet, probably far more, all lying in one undisturbed
mass. It is suggestive that all the glacial deposits
which I have met with in Arctic and Polar lands,
with the exception of the terminal moraines now
forming above sea-level in areas so widely separated
as Smith's Sound, Grinnell Land, North Greenland,
Spitsbergen, Novaya Zemlya, and Arctic Norway,
should be glacio-marine beds. Throughout this
broad expanse of the Arctic Regions I have come
across no beds that could be satisfactorily assigned
to the direct action of land-ice; that is to say, beds
formed *in situ* by the grinding force and pressure of
an ice-sheet. On the contrary, so far as I can judge,
the glacial beds which I have traced over the extensive area mentioned above have all been deposited
subaqueously and re-elevated."

One of the strongest arguments that can be used
against the view of the marine origin of the glacial
phenomena in Northern Europe seems to me the fact
that we find polished rock-surfaces far removed from
the source of glaciers, and so exactly resembling those
produced at the present day by our Alpine glaciers as
to appear identical to the experienced eye. Most of
such striated and polished rocks occurring in the
higher mountain ranges of Scandinavia, and also of
the British Islands, have no doubt been actually produced by glaciers, whilst those in the plain, sometimes hundreds of miles away from the mountains,

must have originated in a similar manner; that is to say, by a heavy mass of material containing stones being slowly dragged over the rock-surfaces. The weight which causes the stones to polish the latter is generally ice, but it is quite conceivable that any other substance, especially if it is in a semi-solid state, must act and operate in much the same way. All polished rock-surfaces are carved by glaciers, because we can see them done by glaciers every day, is the argument commonly used nowadays. It was not so formerly. But Mr. Mallet and his views are almost forgotten now; his name does not even appear in our great modern works on the Ice Age. His argument was that as the land rose out of the glacial sea, the mud which had accumulated round the shore slipped downward in a direction determined by the contour of the surrounding valleys and mountains. The moment the land rose above water-level, the large mass of gravel and mud lying upon it slipped downward. During a steady rising of the land there would therefore be produced a continuous sliding down of this mud-glacier, which would groove and polish the rock underneath it, in the same manner as the ice-glaciers do in the Alps (p. 47). Professors Sedgwick and Haughton became strong adherents of Mr. Mallet's theory at the time, but it seems later on to have fallen into disfavour with geologists, who may not even be thankful to have it brought to light again.

SUMMARY OF CHAPTER II.

I have endeavoured to show in this chapter how we can determine approximately the original home of an animal. By this means we are able to study the component elements of the European fauna, which is found to consist to a large extent of migrants from the neighbouring continents. There is a Siberian, an Oriental, and an Arctic element in it. The remainder of the fauna is derived from local centres of dispersal. What was formerly believed to have been one great northern migration now resolves itself, on closer study, into two very distinct ones—the Siberian and the Arctic. The mammals have received most attention hitherto, because their remains are so frequently met with, thus enabling us more easily to investigate their past history; but butterflies and snails have not been neglected, and at least one very remarkable work on the latter has been published dealing with their origin in Europe and in the remainder of the Palæarctic region.

The former distribution of land and water is intimately connected with the origin of the European fauna, and the changes which have taken place in this respect may be best traced by the present distribution of mammals, snails, and earthworms. In this manner the British Islands may be shown to have been connected with one another and with the Continent; Spain with Morocco across the Straits of Gibraltar; Greece with Asia Minor, and so forth.

The British fauna has played such an important part in the evolution of the European fauna, that it forms the key to the solution of the wider problem. In it five elements are recognisable, of which the Lusitanian element is the oldest, and the Siberian the most recent. It has been deemed advisable to conclude this chapter with a short review of the

history of the Glacial period in its climatic effects on the
animals and plants of Europe. A number of writers are
quoted who have conducted special researches in determining
the temperature of our continent at the time. The fauna of
Europe is frequently described as having been of an Arctic
nature, but as a matter of fact there existed during the Ice
Age a striking and most remarkable mingling of a northern
and a southern fauna. The presence of Siberian mammals in
Europe is said to have been due to the prevalence of a dry
steppe climate, but this view is not supported by other evidence.
The Alpine flora in a wide sense is probably pre-glacial in
origin, and appears to have survived the Ice Age where it
is now known to exist. A few words on the phenomena of
glaciation are added before bringing the chapter to a close.

CHAPTER III.

THE British Islands are, as I have remarked, very suitable as a starting-point for our investigations. Their fauna and flora are fairly well known, and the distribution of the large animals at any rate, which are of course of much importance in these researches, has been as much studied as that of any other area in Europe. We possess in England an abundance of the remains of past animal life, and a combination of the data furnished by both of these important factors will enable us to draw up a history of the origin of the present British fauna.

In the first chapter I indicated that in the fauna of the British Islands three divisions or elements are recognisable—a northern, a southern, and an eastern. These elements correspond to migrations which can be proved to have arrived in this country at different periods in past times. When we investigate these migrations more closely, the eastern is found to be composed partly of European and partly of Siberian species. The southern is made up of European and of Central and Southern Asiatic species. To make matters still more complex, the southern and

89

eastern migrations insensibly merge into one another, so that it is often very difficult to determine to which of them an animal may belong. The European species spread principally from three centres över Europe—viz., from the Lusitanian, Alpine, and the Balkan centres. The southern element of the British fauna is therefore composed of animals which have originated in these three centres, and in Central and Southern Asia. The Balkan species have been included with those coming from the latter centre under the term "Oriental" migration. The sixth chapter is devoted to it, whilst the Lusitanian and Alpine migrations have each a chapter to themselves.

The Arctic Hare is, as I have already mentioned, one of the mammals of the northern element of the British fauna. It is now confined to the mountains of Scotland and the plain and mountains of Ireland. But in former times it had a wider range in the British Islands. The Stoat is another distinctly northern mammal. It occurs with us, as Messrs. Thomas and Barrett-Hamilton have pointed out, in two distinct varieties or species, the one being confined to Great Britain, the other to Ireland. As I shall explain more fully later on (p. 135), I have reasons to believe that the Irish Stoat came from the Arctic Regions as a northern migrant, but that the English Stoat, on the other hand, reached England with the Siberian fauna from the east. A third northern animal, now extinct in the British Islands, is the Reindeer. It is supposed to have died out in these countries not very many centuries ago, and records have been handed down to us that it still inhabited Scotland as late as the thirteenth century. Like the Stoat, it occurred in two well-known varieties, distinguished from one another by the shape and form of the antlers. In the English pleistocene deposits the remains of both kinds are met with mingled together, whilst in Ireland only one of them has been found. The explanation of this case is similar to that of the two stoats. One of the varieties, which we may call the northern one, came to us from the Arctic Regions; the second wandered to the British Islands at a later period, when Ireland had probably become separated from England. It was therefore unable to penetrate so far west.

One of the most familiar examples of a northern British bird is the Red Grouse (*Lagopus scoticus*). By most authorities it is looked upon as a species distinct from the Scandinavian Willow Grouse (*Lagopus albus*), but except in colour it is undistinguishable from it, and the eggs are identical. The whole genus *Lagopus* is a distinctly Arctic one, and there can be no doubt that the British Grouse belongs to the northern migration, just like the Arctic Hare. The Ptarmigan (*Lagopus mutus*) and the Snow Bunting are also migrants from the north. Though as resident British birds they are quite confined to Scotland, the remains of the former have been found in a cave in the south of Ireland, showing that its range in the

British Islands was formerly more extensive. Another bird which probably came to our shores with this same migration, though it is now unfortunately extinct, is the Great Auk (*Alca impennis*), of which some specimens have luckily been preserved in our museums. From the occurrence of its remains in kitchen-middens and other recent deposits, the Great Auk is known to have inhabited the coasts of Scotland, Ireland, and Scandinavia, as well as those of Newfoundland. Mr. Ussher recently found the bones of this bird near Waterford, which, I believe, is the most southern locality known. The manner of their occurrence leaves no doubt that the bird had been used as food by the early races of man. In all probability it originated in the Arctic Regions, and subsequently spread south on either side of the Atlantic. We need not here refer to the many winter visitants, —northern birds which appear regularly, or at more or less long intervals, in these islands,—although in most of the ornithological works they are included under the term " British Birds."

All the British reptiles and amphibia appear to have reached us from the south or east, but among the fishes there are a good many northern forms. The whole salmon family — the *Salmonidæ* — are typical northern immigrants. The Stickleback (*Gasterosteus aculeatus*), too, has undoubtedly come to us from the north. The genus *Cottus*, like *Gasterosteus*, is certainly Arctic in origin. Originally freshwater forms, many species are now found between tidemarks, and of these a few have migrated southward along the coasts of the great continents. Thus we meet with various species of *Cottus* as far south as California and Japan, on the American and Asiatic coasts of the Pacific respectively. In Europe, two species, viz., *C. scorpio* and *C. bubalis*, range as far south as the French coast. Our freshwater *Cottus*, the Miller's thumb (*Cottus gobio*), has migrated to us from the north with the Arctic species. All the freshwater forms, indeed, of this genus are typically Arctic.

A large number of land and freshwater invertebrates too have no doubt reached us from the north. Some of them may have originated in Scandinavia or within the Arctic Circle, but others probably came still farther, either from America or even from Asia, and used the Arctic land-connection *viâ* Greenland in their migration to Europe. As I shall give a number of additional instances of such migrants in the succeeding chapters, I need not, perhaps, dwell upon them now any longer, except to mention a few of the more typical ones. *Vertigo alpestris*, a minute snail with an amber-coloured shell, and our freshwater pearl-mussel, *Unio (Margaritana) margaritifer*, belong to this migration. Then among butterflies we may cite the Marsh-ringlet (*Coenonympha typhon*), and among beetles, *Pelophila borealis* and *Blethisa multipunctata*. There are a number of northern spiders, among which a few certainly indicate an Arctic origin, or at any rate, that they have wandered to

Europe across Greenland and the old Arctic land-connections. *Bathyphantes nigrinus, Linyphia insignis,* and *Drapetisca socialis,* for instance, are three British species whose range indicates a northern origin, and which also occur, according to Mr. Carpenter, in North America. Mr. Carpenter also tells me that the Collembolan, *Isotoma littoralis,* is a typical northern migrant. He has recently discovered it in the west of Ireland, its only station in the British Islands.

Among the crustacea, the genus *Apus* forms an exceedingly interesting illustration of the northern migration, *Apus glacialis* having been discovered in a Scottish pleistocene freshwater deposit, whilst it is now almost confined to the Arctic regions.

To the same group of animals also belong the three remarkable species of freshwater sponges, *Ephydatia crateriformis, Heteromeyenia Ryderi,* and *Tubella pensylvanica,* which Dr. Hanitsch has described from some lakes in Western Ireland. None of these are known from Great Britain or from the continent of Europe. A few North American plants grow wild in the same district. That any of these should owe their existence in Ireland to accidental introduction appears to me exceedingly improbable. In a former contribution to this subject (*a,* p. 475) I assumed that these American plants and animals had migrated to Europe at the same time as the other northern forms referred to. My friend Mr. Carpenter, however, takes exception to this (p. 383), and I quite recognise the force of his argument. "Their very restricted and discontinuous ranges," he says, "along the extreme western margin of Europe mark them as decidedly older than those northern animals and plants which have a circumpolar distribution." We have indeed quite similar examples in the Oriental migration, of which part is very ancient, surviving here and there and exhibiting discontinuous distribution. We may therefore look upon these American immigrants as among the oldest members of that northern stock which have survived in our islands —probably a mere remnant of a once luxuriant flora and fauna.

In order to show the importance of the Eastern or Siberian element in the English, or, we might say with Dr. Sclater, the Anglo-Scotian mammalian fauna, I herewith give a list of the species of mammals which probably migrated to Great Britain from Siberia. I have marked with an asterisk those which still exist in this country (not in Ireland), or have become extinct within historic times.

Canis lagopus.	* Mus minutus.
Gulo luscus.	* Arvicola agrestis.
* Mustela erminea.	* „ amphibius.
* „ putorius.	„ arvalis.
* „ vulgaris.	* „ glareolus.
* Sorex vulgaris.	„ gregalis.
Lagomys pusillus.	„ ratticeps.
* Castor fiber.	Equus caballus.
Spermophilus Eversmanni.	Saiga tartarica.
„ erythrogenoides.	Ovibos moschatus.

Cricetus songarus.   Alces latifrons.

Myodes lemmus.   „   machlis.

Cuniculus torquatus.   Rangifer tarandus.

We have evidence that most of these twenty-six species of mammals came from Eastern Europe, but there is no reason to suppose that they originated there. On the contrary, it is highly probable, as I said before, that their native home is Siberia, and that they entered Europe to the north of the Caspian. Along with these, vast numbers of other forms of life, and also plants, swarmed into our continent, and as we advance eastward from England we meet with them in increasing numbers to the present day. But not only on the Continent do we find these survivals of the great Siberian migration, which has been so ably described by Professor Nehring; no less than nine species still inhabit Great Britain (if we include the recently extinct Beaver). On the other hand, not more than three have been found fossil in Ireland, and of these only one still survives. This very significant fact will be referred to again more fully on p. 153. Meanwhile it should be remembered that these three species, viz., *Mustela erminea*, *Equus caballus*, and *Rangifer tarandus*, occur in Ireland in varieties distinct from those found in Central Europe. It is upon this, and many other circumstances, that I founded my belief that Ireland was already separated from England at the time of the arrival of the Siberian emigrants in the latter country. As we shall see, the Irish Stoat, Horse, and Reindeer probably came by a different route from that taken by the English representatives of the same species.

Very few of the lower animals of Siberian origin have reached the British Islands. Most of those which were formerly thought to be Siberian are either of East European or of Central and South Asiatic origin, though they probably joined the Siberian migration on their way to England. The Arctic migration brought a greater variety of species to this country than the Siberian, but neither the one nor the other has contributed more than a small percentage to the British fauna. The bulk of that fauna is derived from the various European centres of dispersal, and especially from Central and Southern Asia.

Those animals which have their home in the latter area, I have named Orientals, though it must be remembered that they need not necessarily have come from what is known among zoologists as the "Oriental Region." The terms "Oriental animals" and "Oriental migration" are used here in a wider sense, and include even those species which reached Central and Northern Europe from South-Eastern Europe. It is astonishing, what a vast number of both vertebrate and invertebrate animals can be traced back to this Oriental migration. Great tracts of Europe were repeatedly submerged beneath the sea during Tertiary times, and on their re-appearance were formed into green fields and pastures new for the rich Asiatic fauna, which was ever ready to flood the neighbouring

continent. This went on, and not for a comparatively
short space of time, as in the case of the Siberian
invasion ; the immeasurable ages which passed,
whilst several of the Tertiary epochs dawned upon
Europe, witnessed an almost constant stream of
Asiatic immigrants pouring in upon us. Europe
returned her own products in exchange, but they
must have been scanty in comparison to the enormous number of species which have undoubtedly
originated in Central and Southern Asia. Very
many of the widely distributed forms in the British
Islands are of Oriental origin. Among these are
also the cosmopolitan species, such as the Barn
Owl (*Strix flammea*) and the Painted Lady Butterfly (*Vanessa cardui*). A great number of our
British Mammals, Birds, Butterflies, and Beetles have
come to us with the Oriental migration. But, as
I shall explain in the special chapter devoted to it,
the earlier migrants from the south-east found their
northward progress barred by a great sea which
stretched through Central Europe from west to east.
The Mediterranean was then divided into two smaller
basins. On their arrival in Greece, which was then
connected with Asia Minor and Southern Italy,
the Oriental migrants seem to have turned westward,
skirting the shores of the Mediterranean. When they
finally reached Spain, many then changed their course
northward (see Fig. 5, p. 117) and wandered to the
British Islands with the Lusitanian animals which
came from South-Western Europe.

Dr. Wallace makes mention of a fairly large number of species and varieties of Lepidoptera, Coleoptera, and land and freshwater Mollusca, supposed to be *peculiar to the British Islands.* Even if these were all found to be of British origin, most of their nearest relatives are continental species. Many, however, must be looked upon as mere races or sub-species of familiar continental forms. But others, such as *Geomalacus maculosus* and *Asiminea Grayana,* are by no means confined to the British Islands. Some of the so-called varieties enumerated by Dr. Wallace are merely slight individual variations in form and colour, which, only by the extraordinary tendency of the variety-monger to advertise himself, have received a distinct Latin denomination. The number of the remaining species, after weeding out the unworthy ones, will be found to be insignificant.

Similarly, the list of seventy-five species and varieties of flowering plants included by Dr. Wallace among the forms peculiar to the British Islands (p. 360) is reduced by Sir Joseph Hooker to twenty. The remainder are to be considered as varietal forms of a very trifling departure from the type, or as hybrids.

Just as we distinguish in the British Islands the parts inhabited by Englishmen, Scotchmen, and Irishmen, so we can recognise three divisions in the animal world, and these roughly correspond to the boundaries of England, Scotland, and Ireland. Most of the eastern species inhabit England, most of the northern ones are confined to Scotland, whilst Ireland is occupied chiefly by southern animals. This, however, is only a very rough-and-ready method of sub-dividing the British Islands into their component parts according to the origin of their faunas. On closer study such a division is found to be unsatisfactory. The eastern species do not really stop at the Scottish frontier, they range far into Scotland. Nor are the northern forms confined to the latter country. Many of them range into Ireland, and also into England. I have constructed a map of the British Islands showing approximately the boundaries of the northern, eastern, and southern species (p. 7), but even this may not altogether meet with the views of an ornithologist or conchologist. For every group of animals the boundaries would probably require to be marked differently. There is also a good deal of overlapping, so that the attempt to define the limits of the various elements meets with great difficulties. But the map represents, I think, fairly well the general impression one receives as to the disposition of its component elements, after a careful study of the British fauna as a whole.

The distribution of the British plants has been worked out much more thoroughly than that of the animals. It need not surprise us, therefore, that the first attempt to separate the British Islands into natural divisions was made by a botanist—the late Mr. Watson. As he himself pointed out, in making these divisions he did not take into consideration the origin of the British species. They represent merely groups of assemblages of plants of different types of vegetation. Edward Forbes, on the other hand, founded his districts on the origin of plants. His work is not only the first of the kind, but it is a classical essay, and remains one of the most remarkable contributions to the literature on the geographical distribution of living organisms known to science. The vegetation of the British Islands, he informs us (p. 4), presents a union of five well-marked floras, four of which are restricted to definite provinces, whilst the fifth, besides exclusively claiming a great part of the area, overspreads and commingles with all the others. These are—

I. Mountainous districts of South-west and West of Ireland . . } Lusitanian type.

II. South-west of England, and South-east of Ireland . . . } Gallican type.

III. South-east of England.

IV. Mountains of Scotland, Cumberland, and Wales . . . . . } Scandinavian type.

V. General Flora . . . . . Germanic type.

Professor Forbes points out, in connection with the plants of the Germanic type, that the fauna accompanying this flora presents the same peculiarities and diminishes westward and to the north. This type includes, therefore, almost all the species which can be shown to have come to us directly from the east, few if any of which have penetrated to Ireland.

On a previous occasion, the same author had

divided the British Islands into ten districts, according to the distribution of their molluscan fauna. These are—

    I. The Channel Isles.
    II. South-east of England (including Cambridgeshire).
    III. South-west of England.
    IV. North-east of England.
    V. North-west of England (including Isle of Man).
    VI. North of Ireland.
    VII. South of Ireland.
    VIII. South of Scotland.
    IX. North of Scotland.
    X. Shetland Isles.

In a short paper on this subject (*b*, p. 5), I have shown that some of these districts are founded on erroneous data, whilst, with the knowledge now at our disposal, others can no longer be maintained as distinct. I thought then that the molluscan fauna warranted a division of the British Islands into the following two provinces :—

    I. England and Wales (except the South-west).
    II. South-west of England and Wales and the whole of Ireland and Scotland.

The second district contains some species of molluscs which are almost entirely absent from the first, such as *Geomalacus maculosus*, *Testacella Maugei*, *Helix pisana*, *Helix revelata*, *Helix acuta*, and *Pupa ringens*. These are all of Lusitanian origin, and do not occur in Central Europe. Scotland alone cannot be classed as a separate province, since it does not contain a single species peculiar to itself. But, along with Ireland and the South-west of England and Wales, it is distinguished from the remainder of these countries by the almost total absence of what have been called Germanic types.

A French conchologist, the late Dr. Fischer, dealt with the British molluscan fauna in a somewhat similar spirit (p. 57). He divided the British area into two districts, but these differ from mine in so far as the South-west of England and Wales and the West of Ireland form one ; the remainder of England and Ireland as well as the whole of Scotland the other. His classification is of particular interest, since the first district represents part of a larger Atlantic province, the second a portion of the Germanic province of the European sub-region. The latter he looks upon as one of the sub-regions of the great Palæarctic Region. Attention is thus drawn to the intimate relationship existing between the western parts of the British Islands and the Spanish peninsula on the one hand, and between the eastern portions and Central Europe on the other.

Mr. Jordan's North-Sea-and-Baltic district includes Scotland and the North of Ireland, whilst England joined with the West and South of Ireland forms part of his Celtic province. Both of these districts or provinces belong to Mr. Jordan's greater Germanic Region (p. 302).

In the collection illustrating the geographical dis-

tribution of animals in the Dublin Museum, the
British species have been grouped into three divisions.
One contains those with a wide range over the British
Islands, another the characteristic forms of the south-
east and lowland districts of Great Britain, and the
third the Irish and the western and highland Anglo-
Scotian species. Mr. Carpenter has named the last
two divisions the "*Teutonic*" and the "*Celtic*." More
recently, he has recognised that this last division
contains two distinct groups ; one including animals
of northern, the other those of southern origin.
He acknowledges indeed, just as I do, three distinct
faunas in the British Islands, with the addition of
the group of generally distributed species of un-
determined origin.

Many other naturalists have worked in the direction I have indicated—namely, in grouping the
British animals into several distinct assemblages,
without, however, taking their foreign range into
consideration, or their origin. I have already referred
to the useful work done by botanists, who have been
the pioneers in the science of the geographical distribution of living organisms. Among the British
naturalists who have applied the principles of Watson
to zoology, A. G. More deserves to be specially mentioned. He was the first to make a serious study of
the British fauna on the lines laid down by that distinguished botanist. In conjunction with E. Boyd,
he published a valuable essay on the "Distribution of
Butterflies in Great Britain," and later on the birds

were similarly dealt with. All the more important groups of animals are now being studied with a view to determining their exact range in these islands. Mr. Harvie-Brown, Mr. J. W. Taylor, Mr. Eagle Clarke, Mr. Miller Christy, Mr. Ussher, Mr. Barrington, and a number of others have considerably advanced our knowledge in this direction in recent years. Any such contributions are to be welcomed as furnishing us with the necessary data to solve the problem of the origin of the British fauna. Meanwhile we know enough to enable us to assert positively that the latter has reached us by land-connections from various parts of Europe (cf. p. 35). This statement of course refers to the bulk of the British fauna. The small proportion of indigenous species, or such as have been introduced accidentally, may be left out of consideration when dealing with the great mass of animals which have evidently migrated to the British Islands on land now sunk beneath the sea (see Fig. 4, p. 60). Opinions of zoologists, botanists, and geologists are practically unanimous on this subject; yet there are two other theories, which have from time to time been advanced to account for the origin of the British fauna. Only the first of these, however, can claim the serious attention of those interested in the problem. Its chief contention lies in the oft-asserted dictum of the "*imperfection of geological record.*" It has been suggested, in fact, that the British fauna, instead of having migrated to our islands, might have originated there, but that, owing to the fragmentary nature of our Tertiary deposits, all trace of their early history had disappeared. "The origin of European species," remarks Professor Cole (p. 238), "within the area of the British Isles, and their migration outwards when local conditions became less favourable for their multiplication, are possibilities that seem too often disregarded. Yet the geologist must see in the western borderland of modern Europe a diminished continent from which land-animals must have again and again moved eastward." "Hence geologists may fairly be unwilling to look on our isles as barren lands waiting to be peopled in pliocene or later times. Far rather has the breaking up of a broad land-area along the present continental edge sent our land-fauna to the new steppes that opened eastward, leaving us a mere diminished remnant to struggle with the glacial period."

There are in Professor Cole's views many points with which I readily agree. In the first place, he acknowledges that migration has taken place on land, so that we have our land-connection between Great Britain and the Continent whatever theory we accept as to the direction taken by the migrants. That the western borderland of Europe has given rise to many important assemblages of animals in past times, seems to me also exceedingly probable, nor do I look upon the British Islands as "barren lands waiting to be peopled in pliocene or later

times." On the contrary, I believe an almost uninterrupted stream of migrants poured into the
British Isles before pliocene times from the south.
But what I thoroughly disagree with, is the remark
that our British land-fauna has been sent to the new
steppes that opened eastward. These are the more
or less arid portions of Eastern Europe. Professor
Cole no doubt has in mind those species of mammals
which I have included in what I called the Siberian
migration, and of which we have fossil evidence
in the late Tertiary deposits of Europe. It would
be impossible here to discuss this subject fully,
especially as I have done so in the subsequent
chapters; but, even if we had no geological record
whatsoever, the present range of the species in
question and their nearest relatives must convince
us that they could not have originated in Western
Europe. However, on the strength of the geological
evidence, Professor Nehring—the only one who has
made this fauna his special study—remarks (p. 228),
that there seems scarcely any doubt that this steppe-
fauna just referred to had come to us from the east.
Professors Boyd Dawkins, Brandt, and Lartet held
similar views.

The theory that an ice-sheet stretched across a
narrow sea might be the means of aiding a fauna
across from the mainland to an island, is particularly
inapplicable to the British Islands. Neither Mr.
Kinahan nor Mr. Lamplugh, the two supporters of
this view, have, however, taken the trouble to apply

it to more than one species of the British fauna
ice-bridge, thinks Mr. Kinahan, "could easily
connected Scotland and Ireland, thus giving a
causeway for migration" (p. 3). Mr. Lam
throws more light on this interesting speculat.
giving us the name of an animal which he be
crossed a narrow sea on a bridge of ice. This a
unfortunately happens to be one whose remains
never been found in high northern latitudes, vi
Irish elk (*Cervus giganteus*). And because he
opinion that this species of extinct deer fou
way to the Isle of Man from the mainland
waning ice-sheet, he sees no reason why c
elements of the Irish fauna should not have
similarly introduced.

It seems of no advantage to begin the disc
on the origin of the British fauna by assumin
former existence of ice-bridges, and the possibi
a migration across them of some of its member
a glacier connected Scotland and Ireland, the c
of both countries (since they were highland
acted as the feeders of the ice-sheet) must
been uncomfortable to the majority of the 1
species. What were the inducements that
have prompted those which had braved the
comforts of Scotland to emigrate to Irela:
such a time? What light does it throw o
origin of the Irish fauna as a whole, to advan
extremely improbable hypothesis that certai
ments of it may have reached Ireland by a

bridge? If any species came to that country in such
an unusual manner, surely they must have been Arctic
or northern forms. But what about the southern
species, which form the bulk of the Irish fauna and
also the flora? Even the Arctic element of the
British fauna, which probably includes, besides the
Reindeer, many hundreds of species, could not, I think,
have migrated to these islands on an ice-bridge. Indeed, I agree with most of the writers who have dealt
with the subject, in asserting that the northern as well
as all the other elements of our fauna utilised for their
migration the old land-bridges which connected these
islands with one another and with the Continent.

There is a greater diversity of opinion as to the age
during which the British fauna arrived in these islands.
This is naturally a much more complicated problem,
but it is one which I am convinced will ultimately be
solved mainly by means of a study of the geographical
distribution of animals and plants. If we can settle
the relative ages of the various migrations, we thereby
supply an important link in our attempt to reconstruct
the past geographical features of the British Islands.
The range of the British species will give us an
idea of the nature of the land-connections and their
gradual changes in course of time. Geological data
are exceedingly valuable in these inquiries, but it is
a fatal mistake to build our geographical theories
and the origin of the British fauna as a whole
entirely on the assumptions of a certain school of
geologists. Unfortunately, Dr. White's very interest-

ing remarks on the British fauna for this reason lose much of the value which they might otherwise possess.

In his remarkable essay the late Edward Forbes affirms that the flora peculiar to the west of Ireland, of which the strawberry tree (*Arbutus unedo*) is the most striking example, and which exhibits such strong southern affinities, is not only much the most ancient of our island floras, but that it is actually of miocene age. It migrated to Ireland from Spain at a very remote period, during which he supposed that a direct land-connection existed between the two countries. The destruction of this old land-bridge, he thinks, must have taken place before the commencement of the Glacial period. Climatal changes during that time destroyed the mass of the southern flora which had thus reached Ireland, the survivors being species such as were most hardy (saxifrages, heaths, etc.), which he considers to be the only relics of this most ancient portion of our flora.

The northern or Arctic fauna and flora, according to the same author, established themselves in the British Isles during the Glacial period—at a time when these were groups of islands in the midst of an ice-bound sea. Finally, the great mass of our animals and plants migrated from the Continent to England after the Glacial period. "The migration of the species," he says, "less speedy of diffusion, which are now peculiar to England was arrested by the breaking up of the land-connection between England

and Ireland, and thence the famous deficiencies of the sister isle, as, for instance, its freedom from reptiles" (p. 10). He is also of opinion, that the separation between England and the Continent took place at a later date than that between England and Ireland.

According to Dr. A. R. Wallace (p. 338), we possessed just before and during the Glacial period "a fauna almost or quite identical with that of adjacent parts of the Continent, and equally rich in species." But the submersion, he thinks, which is supposed to have occurred during the latter part of the Glacial period, destroyed the greater part of the life of our country. When England again became continental, continues Dr. Wallace, this fauna was succeeded by an assemblage of animals from Central Europe. "But sufficient time does not seem to have elapsed for the migration to have been completed before subsidence again occurred, cutting off the further influx of purely terrestrial animals, and leaving us without the number of species which our favourable climate and varied surface entitle us to." The comparative zoological poverty of Ireland he attributes to the fact that "the depth of the Irish Sea being somewhat greater than that of the German Ocean, the connecting land would there probably be of small extent and of less duration, thus offering an additional barrier to migration."

Dr. Wallace's explanation of the origin of the British fauna is disappointing after Forbes's careful study and critical inquiry into its component elements. So great an authority on geographical distribution might have given us more lucid statements of his views on a variety of topics connected with this subject.

In speaking of the fauna of Ireland, Professor Leith Adams, Professor Dawkins, and Mr. Alston are evidently only thinking of the mammals, which form but a very small proportion of it. The first-mentioned palæontologist held that there was a land-communication between Scotland and Ireland at the close of the Glacial period, by which the greater portion of the mammals that had found their way to the former country crossed to the latter (p. 100). And, he continues, the severance between the two countries must have taken place before the slow-travelling Mole, the Beaver, the forest-haunting Elk and the Roebuck had time to arrive.

Much in the same spirit are Mr. Alston's remarks on this subject (p. 5). " The absence from the known fossil fauna of Scotland and Ireland of most of the characteristic post-pliocene English animals, shows that the northward migration of these forms was slow, gradually advancing as the glacial conditions of the northern parts of our islands decreased in intensity. Thus it is not difficult to suppose that the Hedgehog, Ermine, Badger, Squirrel, and Mountain Hare may have found their way through southern Scotland into Ireland long before they were able to penetrate into the still sub-arctic regions of the Highlands. Subsequently, when the improvement of the climate had continued, the Shrews and Voles may well have found their way northward along the comparatively genial coasts, before the larger beasts of prey could find a sufficient stock of game."

That the Bear, Wolf, Stag, Horse, Mammoth, and Reindeer lived in Ireland before the Glacial period is considered highly probable by Professor Boyd Dawkins (*a*, p. 152).

Only the Butterflies are dealt with in Dr. Buchanan White's clever little essay on distribution. And, as I remarked before, his conclusions are somewhat marred by the unwarrantable assumption that our islands at no distant date were totally destitute of all plant-life, and were therefore uninhabitable by animals. But his paper differs in so far from most of the others, that he has made a thorough study of the one group he deals with. In some respects it may serve as a model to future students in its general treatment of the problem he has set himself to work out. He adopts the principle, even for butterflies, that though it is possible for them to be blown over from the Continent, they have probably migrated with the rest of our indigenous fauna and flora across the dry bed of the German Ocean. His conclusions are that Britain derived its butterfly fauna from continental Europe in post-glacial times, that the Arctic and Alpine species were the first arrivals, and that one part of the Irish species reached Ireland by way of Scotland, another from the south. He assumes, of course, that Great Britain and Ireland were connected at that time.

Within the last few years the spell which has bound naturalists to accept the theory of a total destruction of life during the Glacial period is happily vanishing, and more enlightened views are gaining ground. The Lusitanian species of plants in the west of Ireland, which had already furnished Forbes with an argument in favour of survival, are also regarded by Mr. Bulman as the remnants of a pre-glacial flora which was exterminated everywhere else by the cold (p. 265). This view of the survival of a pre-glacial fauna and flora has since been accepted by Mr. Carpenter, whilst I also have endeavoured to bring fresh evidence into the field in its favour. We both agree with Edward Forbes in considering the Lusitanian element as the oldest section of our fauna and flora, and that it came long before the Glacial period. But we differ somewhat from him, in so far as we do not limit that element to Ireland. It seems also to be represented in South-western England and Wales, though it is there less conspicuous.

This decision as to the relative age of the British South-western fauna has not been arrived at from any geological considerations. The conviction that it must be older than the other sections has been gained solely from a study of the geographical distribution of the species belonging to that fauna. Many of them exhibit what is known as " discontinuous distri-

bution," which zoologists are agreed to regard as a
sign of antiquity. Thus *Geomalacus maculosus*, the
Kerry Slug, is in the British Islands confined to South-
western Ireland (see Fig. 19, p. 300), and on the Continent it is unknown north of North-western Spain.
The Millepede, *Polydesmus gallicus*, has a wider range
in Ireland, and is also known from France and the
Azores. Two Earthworms of the Spanish and
Mediterranean region, viz., *Allolobophora veneta* and
Georgii, have been discovered in Ireland, but are
apparently unknown in England or France; whilst
the Weevil, *Otiorrhynchus auropunctatus*, does not
occur north of the Auvergne Mountains in France
except in Ireland. A very large number of instances
might be mentioned of species found in South-
western Europe, France, the South-west of England
and Ireland. Enough, however, has been said to
show the nature of the fauna, and there is, as
Forbes has pointed out, a corresponding flora.

A great number of the species belonging to the
South-western British element seem to have originated in South-western Europe, or at any rate to have
spread over our continent from that part. Their
home lay therefore probably in a warm, damp
climate, and it seems a reasonable inference to
suppose that they spread north at a time when
the temperature over the British Islands was much
higher than what it is now. Any one familiar with
our Bristle fern, or Killarney fern, as it is called in
Ireland (*Trichomanes radicans*), will readily admit that

it must have come to us at such an epoch. It at once suggests some shady waterfall in a tropical forest, and indeed the home of the genus is South America. It is one of those plants which have evidently migrated to us from South-western Europe, a mere remnant of a once luxuriant flora.

Sir Archibald Geikie tells us (p. 837), and in the main every one agrees with him, that at the beginning of the Tertiary era in which we now live, the climate was of a tropical and subtropical character in Europe. Gradually it became more temperate, and eventually it passed into a phase of extreme cold, but since that time the cold has again gradually diminished. It is quite evident, therefore, that from a purely geological point of view our south-western flora must have migrated northward before the cold came on, and survived in sheltered localities under the influence of the mild coast climate. Some, however, suppose that there occurred a phase of extreme mildness immediately after the Glacial period, and that it was during that time that the Lusitanian fauna and flora became established in the British Islands. To this Professor James Geikie replies (*b*, p. 169), "there are few points we can be more sure of than this, that since the close of the Glacial epoch—since the deposition of the clays with Arctic shells and the Saxicava sands —there have been no great oscillations, but only a gradual amelioration of climate. It is quite impossible to believe that any warm period could have intervened between the last Arctic and the present temperate conditions without leaving some notable evidence in

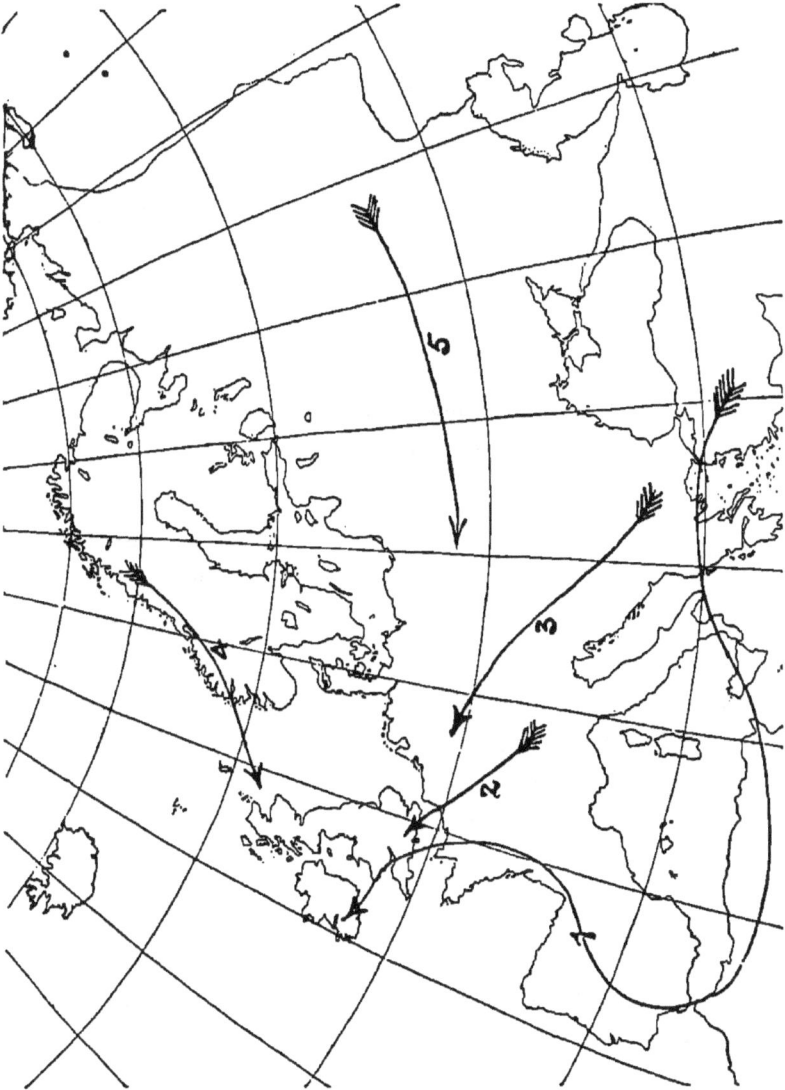

the superficial deposits of Scotland, Scandinavia, and North America." Thus it appears that on the whole

the assumption that the Lusitanian fauna and flora
are very ancient and pre-glacial is also supported
on geological evidence.   ,

The course of events in the origin of the British
fauna might have been therefore somewhat as
follows:—In early Tertiary times, when the climate
all over Western Europe was moist and semi-tropical,
a migration proceeded northward from the south-
western corner of Europe. This was strengthened
by Oriental migrants which had moved westward
along the Mediterranean basin (Fig. 5, No. 1).
Owing to geographical changes supervening, the
Alpine fauna (No. 2) was then enabled to colonise
the British Islands, and subsequently another migration had begun to come in from the south-east
(No. 3). The climate had meanwhile gradually
become more temperate and drier. About the same
time, or even earlier, an Arctic migration commenced
to pass southward (No. 4), and finally the Siberian
animals (No. 5) poured into our continent. The
arrows in the map indicate the directions followed by
the different migrants as they travelled to the British
Islands. The arrows are not meant to represent the
whole nor the full extent of the migrations from
any particular centre, but only in so far as they
affect our islands. Moreover, it would be impossible to indicate on one map the geographical
conditions which obtained during the several migrations. It must be remembered that during the time
which elapsed while they passed into the British

Islands, these were joined in the north to Scandinavia and in the south to Belgium and France. The various phases of geographical evolution of Europe will be studied in the subsequent chapters, and maps will then be given to show as far as possible in a general way the leading characteristics of these great changes.

I have now given some reasons for the belief that several different migrations of animals entered the British Islands in later Tertiary times. I have also shown why some of them must be looked upon as being older than others, and in so far we have come to a decision as to their relative ages. It still remains for us, however, to examine how their geological ages can be approximately determined. We require for this purpose palæontological aid.

In the fifth chapter will be found the history of the Siberian migration. And since we possess most valuable records of it in the numerous fossil remains discovered in Central and Western Europe, we are able to trace their progress from the east to the west in a very complete and satisfactory manner. In England their first appearance dates from the Forest-Bed, for here we find remains of the Glutton (*Gulo luscus*), Musk-Ox (*Ovibos moschatus*), and others (see p. 204). It seems reasonable to suppose, therefore, that the first entry of these Siberian mammals into Europe took place at or just before the Forest-Bed period. But Professor Nehring tells us in his remarkable work on the Tundra and Steppes (p. 222), that in

Germany the remains of the same mammals occur in deposits which are certainly more recent than the lower continental boulder clay; and he is inclined to the belief that they migrated into Europe during the inter-glacial phase which is supposed to have separated the earlier from the later stage of the Glacial period. It is evident that in this case the inter-glacial period in Germany would have corresponded to, and be contemporaneous with, our Forest-Bed period. The deposits immediately preceding the Forest-Bed would also be contemporaneous with the lower continental boulder clay. Although this may seem rather a startling statement to make, from the evidence which will be brought forward in the fourth and fifth chapters I am inclined to the belief that such is probably the case.

Having once arrived at a determination of the exact geological period during which the Siberian mammals invaded our continent, and having also previously determined the relative ages of the various other migrations, we have advanced another step in the direction we are aiming at. Let us suppose that the Siberian migration actually reached the British Islands during the Forest-Bed period. Since the Siberian migration is the most recent of those which entered the British Islands, the others must have commenced their march before the Forest-Bed period. Now it was Professor Boyd Dawkins who first indicated to us, as I have remarked before, the method of research to be adopted in an attempt to determine the geological age of the different migrations in so far as they affected the British Islands. I may be excused, therefore, for again quoting the following important passage in one of his works. "The absence," he says (*b*, p. xxix), "of the beaver and the dormouse from Ireland must be due to the existence of some barrier to their westward migration from the adjacent mainland, and the fact that the Alpine hare is indigenous, while the common hare is absent, implies that, so far as relates to the former animal, the barrier did not exist." The Beaver, Dormouse, and Common Hare are either Siberians or later migrants from elsewhere, and there can be no doubt that at the Forest-Bed period Ireland was already, or was just being, separated from England. All the southern species, that is to say all the Lusitanian, Alpine, and Oriental forms occurring in Ireland, must therefore be older than that period. I have advocated similar views in a former essay on this subject. Mr. Carpenter recently advanced some interesting and valuable criticisms on these views, which we may examine a little more closely (p. 385). "While, then," he remarks, "I find myself in almost complete agreement with Dr. Scharff with regard to the older sections of our fauna, I think that those widespread species which survived the Glacial period must have been confined to the more southern parts of our area, and have only subsequently spread northwards and westwards to Scotland and Ireland." He suggests, in fact, that the widespread British

species belong to a younger or newer section of our
fauna than the local ones. In many cases this may
be quite true, but we possess also a large number .of
common and widely-spread forms which bear the
impress of antiquity upon them. We have the most
positive proof of the antiquity of the very common
small circular Snail (*Helix rotundata*), since it was
found in miocene freshwater deposits near Bordeaux. Many other examples might be mentioned
to show that, though discontinuous range is generally
a proof of antiquity, continuous range is not always a
sign of the opposite. Some species, in fact, appear to
be short-lived and disinclined to spread, whilst others
multiply rapidly even under a change of temperature
and climate, and are to be found almost everywhere.
But even if we supposed, with Mr. Carpenter, that
these widely-ranging species must have been confined
during the Glacial period to the more southern parts
of England, the idea that they afterwards made their
way northwards along the eastern shore of the Irish
Sea and then passed into Ireland, does not appeal to
me. Southern England was occupied at that very
same time by an assemblage of Siberian mammals.
Mr. Carpenter thinks these might have been kept
out of Ireland by an arm of the sea until the land-
connection with North-western England had broken
down. But if an arm of the sea could keep out the
Siberian mammals it would also keep out the widely-
spread British species of the general fauna. On the
other hand, I quite admit that my view of the survival

in Ireland of the pre-glacial fauna is somewhat
difficult to accept, considering that we have such
undoubted evidence of a very extensive submergence.
The case of Isle of Man, quoted by Mr. Carpenter,
can be met, I think, by the supposition that it was
connected with Cumberland until quite recently, and
quite independently of any connection between England and Ireland; that the Isle of Man, in fact,
was always a cape or peninsula of the mainland, and
only recently became separated by local subsidences
or by the action of the sea.

Part of the history of the British fauna will be
referred to again in the next chapter, which deals
with the Arctic migration. We need not therefore
dwell any longer on this subject here. There is one
matter, however, which is of importance in connection
with the geographical conditions of the British Islands
at the time when the greater portion of our fauna
arrived from abroad.

On page 60 will be found a map indicating the
physical geography of that part of the ancient continent on which what are now the British Islands were
situated. Only one large river has been marked on
that map, namely, that flowing out of a lake which
occupied part of the Irish Sea. Another probably discharged its waters into the Atlantic midway between
France and England, whilst the Thames may have
been a tributary of the Rhine, as it emptied itself
into the sea near our south-east coast. I have shown
in a previous essay that the former presence of a fresh-

water lake between England and Ireland is indicated by the distribution of the Charrs and also by the various species of British Coregonus. There are three British species of Coregonus, viz., *C. clupeoides*, *C. vandesius*, and *C. pollan*. These are confined to the lakes of North Wales, North-western England, South-western Scotland, and Ireland. All but the latter communicate at present directly with the Irish Sea. The lakes of the latter country, however, must have done so at a time when the west of Ireland stood at a higher level than it does now. The ancestors of the three Coregonus species, and also those of the Charrs, then lived in the large freshwater lake indicated on the map (p. 60), and when the sea gradually crept up the river valley and finally converted the lake into a gulf, the freshwater fish took refuge in the rivers which supplied it with water.

Now as for the continuous sea-shore between the coast of Brittany and the south-west of Ireland, zoological distribution again aids us in proving that such must have actually existed at no very distant geological date. Most of our common shore forms of life migrate along the coast exactly as land animals do—step by step. Their eggs are carefully attached to fixed objects, so as not to be carried away by the waves, whilst the young often remain and grow old in some particular little pool, rarely venturing farther than a few yards from the spot where they first saw the light of day. A number of such shore forms are found on the west coast of France, the same species recurring again on the south-west coasts of England and Ireland, thus clearly indicating a former continuity of coast-line between these points, now separated by deep sea. A very familiar example to British zoologists is the purple rock-boring Sea-urchin (*Strongylocentrotus lividus*), but there are a great many others, such as the semi-marine Beetles *Octhebius Lejolisii* and *Æpophilus Bonnairei*, the Crustaceans *Achæus Cranchii*, *Inachus leptochirus*, *Gonoplax angulata*, *Thia assidua*, *Callianassa subterranea*, the Fishes *Blennius galerita* and *Lepadogaster Decandollii*, and the Molluscs *Otina otis*, *Donax politus*, and *Amphidesma castaneum*.

Before concluding this chapter, a few words as to my views on the conditions prevailing during the Glacial period will not be out of place. They do not differ very much from those held formerly by most geologists; and even at present there are, as I have mentioned before, a few upholders of those older views.

The sea, I think, must have gradually crept across England from the east during, or shortly after, the Forest-Bed period, so as to separate the south from the north, whilst Ireland and Scotland were then still connected with one another. At a later stage, the sea also partially invaded Ireland, and this condition is very roughly represented on the accompanying map. Mr. Kendall kindly drew my attention to the fact that several notable areas on which shelly drift has been observed are here placed upon the land; but it must be remembered that one stage only can be shown on the map, and that the sea covered more ground a little later. Many of the smaller islands in the glacial sea, too, are not shown. The map, in fact,

FIG. 6.—Map of the British Islands, showing approximately in what manner the sea may have invaded the country from the east during, or shortly after, the Forest-Bed period. The darkly shaded parts indicate the areas covered by water, and the lightly shaded and white portions what was land at that time.

is merely meant to give a general idea of the manner in which the great northern sea moved westward and slowly covered a large portion of the British Islands. These peculiar geographical conditions explain, I think, better than anything, the absence from parts of the Midlands and the north of England of such a number of terrestrial invertebrates which are otherwise widely distributed over the British Islands. In spite of the fact that a large portion of the British Islands became submerged, we possessed at that time an extensive area which has since been claimed by the sea, so that there was ample room for the present fauna to survive the Glacial period. The climate during this period was probably much the same as it is at present, though moister, with cooler summers and milder winters.

It may be asked what proof we have of such an extensive submergence of England and Ireland. My own views are principally based on the general distribution of the fauna in the British Islands, and the belief that nothing but a mild climate during the Glacial period could have brought it about. On purely geological grounds, however, some geologists, notably Mr. Mellard Reade, have come to a similar conclusion. " The whole of Lancashire and Cheshire," he remarks (*a*, p. 542), "from sea-level up to about 400 feet, and in places 600 feet, is covered by a continuous mantle of boulder-clay and sands." "These clays, as a rule, contain distributed through them, in a greater or less degree, fragments of shells and some perfect ones. I myself have recorded forty-four species." Again he continues (pp. 545 and 546): "A large part of Ayrshire is covered with similar shelly boulder-clays from sea-level to 1061 feet at Dippal. These Ayrshire high-level shells have, in the majority of cases, been taken, not from sand and gravel beds, but from boulder-clay, and in that respect they are most important and unique. In Moel Tryfan the shells are found in sands and gravels at 982 feet; on the range of hills from Miaera to Llangollen from 1000-1200 feet; also in sands and gravels at Gloppa, near Oswestry, at 1100-1200 feet; and near Macclesfield at a level of about 1200 feet. In Ireland marine shells can be traced almost from sea-level to a height of over 1000 feet."

"Again," continues the same author, "if we look broadly at the distribution of these shelly deposits, we find that they occur all round our maritime coasts in Lancashire, Cheshire, and Wales, in Cumberland and Westmoreland, Wigtonshire and Ayrshire, and along the eastern coast of Ireland. The same is to be said of the eastern coasts of England and Scotland."

That a very considerable change of sea-level has taken place in some parts of the British Islands would appear to a zoologist the most logical conclusion after an examination of these "high-level shelly sands and gravels," but the shells contained in them are now generally supposed to have been carried there frozen in the sole of a glacier or pushed up in front of it. The older view, however, which agrees so much better with the facts of distribution, fortunately has not disappeared among geologists. "When we call up," says Mr. Mellard Reade (*b*, p. 435), "before our mental vision the simple and well-known facts of nature which suffice to explain the marine drifts on the theory of submergence, it seems unnecessary to resort to the ingenious and artificial system of physics elaborated to explain the phenomena of land-ice."

"When we have more knowledge of the glaciers of the Arctic Regions, and facts, in place of ingenious suppositions, to base our reasoning upon, we may possibly have to revise all our glacial conceptions. In the meantime, the submergence theory of the origin of high-level shelly gravels and sands seems to me by far the simpler of the two theories, and the most consistent with the facts and phenomena which the labours of a succession of enthusiastic geologists have made us acquainted with."

Among those geologists, and they form the majority, who hold that Ireland was covered by land-ice, there is a great diversity of opinion as to its extent. Messrs. Close, Kinahan, J. Geikie, and others believe that the ice covered practically everything, whilst others who claim to have examined the ground with equal care, such as Professor Carvill Lewis, were led to believe that the south of Ireland, with the exception of a few local glaciers, was free from ice. The glacial

9

phenomena of the country can therefore be interpreted in different ways, even by those who are convinced that they are due to land-ice and not to icebergs or mud-glaciers.

SUMMARY OF CHAPTER III.

The history of the British fauna is not only of interest to us from a sentimental point of view, it is a convenient starting-point in the study of the larger European problem. The fauna, broadly speaking, is composed of three foreign elements, viz., the northern, eastern, and southern, to which may be added a small endemic one. Examples are given of the more noteworthy forms belonging to each of these. This leads us to the subject of the natural divisions of the British Islands according to their animal inhabitants. Zoologists attempted at first to subdivide these countries, on the lines laid down by botanists, into a large number of provinces. Forbes proposed ten such divisions for mollusca, and subsequently five, which were ultimately reduced by others to two or three.

The opinions of biologists are almost unanimous in attributing the bulk of the British fauna and flora to migrations by land from the Continent, but two other theories, viz., those of Professor Cole and Messrs. Kinahan and Lamplugh, are also referred to. The first believes in a possible migration eastward from Western Europe, and the latter support the view of the former existence of ice-bridges to assist the fauna in their migrations.

An endeavour is next made to determine at what geological periods the various migrations entered the British Islands. There is considerable difference of opinion on this subject. Some believe that the British fauna is altogether post-glacial; a few think that it is partly so and the remainder glacial;

others again hold that a portion is pre-glacial and the rest glacial and post-glacial. Those who have studied the subject most closely feel convinced that the south-western or Lusitanian fauna, and also the flora, must have arrived before the Glacial period and survived the latter in these Islands. It seems reasonable to suppose, therefore, that the climate cannot have been very severe during the so-called Ice-Age. This Lusitanian fauna must be looked upon as the oldest portion of the British fauna. The Alpine and Oriental migrations arrived next. After these came the Arctic, and finally the Eastern or Siberian. As the fossil evidence is most complete with regard to the last, we are able to determine with precision not only the direction whence this migration came, but approximately its geological age. It arrived in Germany from the east after the deposition of the lower boulder-clay. Since the boulder-clay is looked upon as a glacial deposit, the Siberian migration reached Central Europe after the first portion of the Glacial period had passed. In England it makes its first appearance in the Forest-Bed, which would therefore correspond to the "Loess" formation of Central Europe. All the other migrations are older than the Siberian. They must therefore have come to Great Britain during the earlier part of the Glacial period or before it.

The chapter concludes with a short statement on the physical geography of the British Islands during the time when these migrations entered them. That there existed a continuous coast-line between France and Ireland is proved by the occurrence of a considerable number of identical shore species, whilst the former existence of a freshwater lake on the site of the present Irish Sea is indicated by the distribution of some freshwater fishes.

CHAPTER IV.

THE ARCTIC FAUNA.

THE lands lying within the Polar Circle are inhabited by an assemblage of animals and plants, many of which are peculiar to those regions. They are mostly adapted to the abnormal conditions of life prevailing in the high latitudes of our globe—the long, dark winters, and the short summers of one long day. Though the numbers of species and of individuals are few, there is a keen struggle for existence in those regions. The prevailing colour of the ground is white, and since a resemblance in the colour of an animal to the ground it lives on acts as a protection to weak ones, and also enables Carnivores to approach their prey with greater facility, it is not surprising that we should find the majority of polar animals coloured white. As I remarked, the polar area contains a very distinct set of species; most of them, however, range beyond the confines of the Arctic Circle. It is therefore scarcely justifiable to raise this Arctic area into a distinct zoological region equivalent to the great zoo-geographic regions, which have been established by Sclater and Wallace, though we might, with Dr. Brauer, look upon it as a sub-region.

There are six typical Polar Land-mammals, one of which, the Polar Bear, is semi-aquatic. The Reindeer (*Rangifer tarandus*) occurs upon almost all the polar lands, and it has often been a source of speculation in what manner it has reached such remote islands as Spitsbergen and Novaya Zemlya—the former of the two being so remote from a continent. There is no doubt that Reindeer are great wanderers, owing to the difficulty of finding sufficient food-supply for the large herds in which they are accustomed to travel; and for this reason they can cross, and have been known to cross, distances of from ten to twenty miles on ice. The Behring Straits, when frozen over in winter, is frequently traversed by them. But I quite agree with Dr. Brauer (p. 260) that it is impossible to account for their presence in Spitsbergen by an immigration from either Novaya Zemlya, Greenland, or Scandinavia, under the present geographical conditions. The seas between the former island and the other land-masses referred to are rarely entirely frozen over. Even if this should occur, the distances between Spitsbergen and Greenland, Novaya Zemlya, or Scandinavia are so great, that a migration across ice is quite excluded from the range of possibilities, since Reindeer could not subsist without food during the time it would take to travel from one to the other. The manner in which it did reach Spitsbergen and Greenland will be discussed more fully below, and I will therefore proceed to mention the other Arctic mammals.

One of the most important and most typical species is the Polar Bear (*Ursus maritimus*), the greater part of whose life is spent on the ice and in the sea. The fact that its favourite nourishment consists of seals proves its excellent and keen faculties of sight and hearing, and its facility in swimming. But it is not a

FIG. 7.—The Musk-Ox (*Ovibos moschatus*). (From Flower & Lydekker's *Mammals*, p. 358. London: Adam & Chas. Black.)

dainty feeder, and lives upon almost all animals which come within its reach ; birds, land-mammals, or fish are not despised in times of scarcity. Its fur throughout the year is coloured white, though in old bears it assumes a more yellowish hue.

Another large mammal, perhaps less well known, is the Musk-Ox (*Ovibos moschatus*, Fig. 7), which resembles in size the smaller varieties of Oxen, but in structure and habits is closely allied to the Sheep. As is implied by the specific name, it exhales a musky odour; this does not, however, appear to be due to the secretion of a special gland, as is the case in other animals with a similar smell. The skin is covered with long brown thickly-matted hair, interspersed with white. It is confined to the most northerly parts of North America and the American Arctic islands, and to North Greenland. Though not now living in the Old World, it seems formerly to have been abundant in Siberia, and, as we shall learn later on, it was one of the species which took part in the great Siberian invasion of Europe. Its remains have been found not only in Germany and France, but also in the south of England.

The Polar Fox (*Canis lagopus*) occurs throughout the Polar Regions, and on islands where even the Reindeer and the Musk-Ox are unknown. Beyond the Polar Circle, its range extends into Northern Asia, to the extreme north of North America, and the mountains of Scandinavia. Like its congeners, it had in pleistocene times a more southerly extension, and fossil remains have been met with in various parts of continental Europe and in England.

The Stoat (*Mustela erminea*), which is known and much valued in commerce under the name of Ermine, was formerly believed to occur only in Arctic America and the northern parts of the Old World, but in more recent years it has been discovered in a number of the northern islands, such as Saghalien, in the islands of the Behring Straits, the Aleutian islands, and also in Greenland and Spitsbergen. In Europe, it is found as far south as the Arctic Hare, or perhaps' even farther, and it flourishes in the Alps up to a height of 9000 feet. It offers a parallel to the Arctic Hare in the fact that in some countries, such as Ireland, it only rarely turns white in winter. The Irish form of the Stoat differs so much from the English, that Messrs. Thomas and Barrett-Hamilton are of opinion that it is specifically distinct, as I mentioned in speaking of the divisions of the British fauna (p. 90).

The Arctic Hare (*Lepus variabilis*) is almost the only one of the typical Arctic mammals which still inhabits the British Islands, and for that reason it is to most of us more familiar than any of the preceding species. Hares have been described from Greenland by the name of *Lepus glacialis*, from the European Alps as *Lepus alpinus*, and under other names from Arctic North America; but though slight differences in the fur and even in the skull can be pointed out, there is no doubt that all these are only varieties or races of what, in the British Islands, is known as the Irish or the Scotch Mountain Hare, *Lepus variabilis*. In the Arctic Regions this Hare remains white throughout the year, but in Scandinavia and some other parts its fur becomes brown in the summer, and in Ireland it frequently remains entirely brown during the whole year, and never, or only in very rare cases, becomes entirely white in winter. Besides Scandinavia, Scotland, and Ireland, it. is found in Northern Russia, and also in the Pyrenees, the Alps, and the Caucasus. In Asia it occurs not only on the mainland of Siberia, but it has

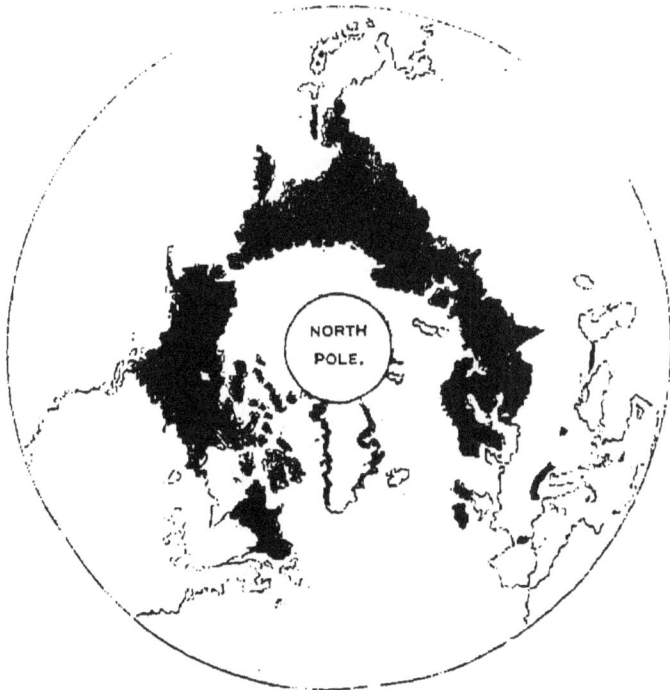

FIG. 8.—Map of the northern hemisphere, to show the geographical distribution of the Arctic Hare (*Lepus variabilis*) indicated in black.

been obtained on the Akita Mountains in Japan and on the Mioko San Mountain, and also on the island of Saghalien. It had in former times a more extensive range, and its remains have been discovered in

England and in a number of places on the continent of Europe. The peculiarity of its range, which will be explained more fully directly, lies in the fact of the occurrence of isolated colonies in the mountains of Europe, in Ireland and Scotland, and in the mountains of Japan (Fig. 8). From a distributional point of view, it is one of the most interesting species of mammals, and its history throws a flood of light on the geographical changes which have occurred in former times.

One more species must be mentioned, and that is the Banded Lemming (*Cuniculus torquatus*), which occurs chiefly in Arctic America, Northern Siberia, and Greenland. Though frequently mistaken for the Scandinavian Lemming, there is a striking difference in the character of the teeth, which has induced zoologists to put them into distinct genera. The Arctic Lemming, moreover, is distinguished from the Scandinavian by the absence of external ears, the densely furred feet, and by the great length of the two middle claws in the fore-feet. There are two species of the true Lemming, namely, the one just referred to, *Myodus lemmus*, and *Myodus obensis*. These may be looked upon as more or less Arctic species, since they occur within the Polar Circle, but they are not so exclusively confined to that region as the Banded Lemming (*Cuniculus torquatus*). The remains of both *Cuniculus torquatus* and of *Myodus lemmus* have been found in British pleistocene deposits.

Until recently no Lemming remains had been found

to the south of France, but Mr. Barrett-Hamilton
announced to us a short time since that Dr. Gadow
had discovered some skeletons with their skins still
preserved in a cave in Northern Portugal. These
were found to belong to the Scandinavian Lemming
(*M. lemmus*), and the author incidentally expressed the
opinion that there was some possibility of this species
still inhabiting the mountains of Spain.

The Lemming multiplies with great rapidity under
favourable conditions. In speaking of his experiences
in Siberia Dr. Brehm says (p. 79): " All the young of
the first litter of the various Lemming females thrive,
and six weeks later at the most these also multiply.
Meanwhile the parents have brought forth a second
and a third litter, and these in their turn bring forth
young. Within three months the heights and low
grounds of the tundra teem with lemmings, just as
our fields do with mice under similar circumstances.
Whichever way we turn we see the busy little crea-
tures, dozens at a single glance, thousands in the course
of an hour. But the countless and still increasing
numbers prove their own destruction. Soon the lean
tundra ceases to afford employment enough for their
greedy teeth. Famine threatens, perhaps actually sets
in. The anxious animals crowd together and begin
their march, hundreds join with hundreds, thousands
with other thousands, the troops become swarms, the
swarms armies. They travel in a definite direction,
at first following old tracks, but soon striking out
new ones; in unending files—defying all computation

—they hasten onwards; over the cliffs they plunge into the water. Thousands fall victims to want and hunger; the army behind streams on over their corpses; hundreds of thousands are drowned in the water or are shattered at the foot of the cliffs; the remainder speed on; other hundreds and thousands fall victims to the voracity of Arctic and red foxes, wolves and gluttons, rough-legged buzzards and ravens, owls and skuas which have followed them; the survivors pay no heed. Where these go, how they end, none can say; but certain it is, that the tundra behind them is as if dead, that a number of years pass ere the few who have remained behind and have managed to survive slowly multiply and visibly re-people their native fields." This eloquent passage reminds us of the manner in which migrations of all kinds of animals have taken place in former times, and are still taking place. It is principally want of food which compels them to search for new homes.

On page 91 I have referred to some birds which have come to us from the north. One of these, the Snow Bunting (*Plectrophenax nivalis*), is a typically Arctic species. In summer it is widely distributed, and is found in Spitsbergen, Novaya Zemlya, Siberia, and the Arctic Regions generally. In winter it migrates down into North America, into Japan, Northern China, Turkestan, Southern Russia, and occasionally even across Europe into North Africa. Very characteristic Arctic birds are the Eider Ducks belonging to the genus *Somateria*. Three species have visited the British Islands. The common Eider Duck (*S. mollissima*), which is of such high commercial value, is abundant in Norway and northward,

FIG. 9.—The Great Auk (*Alca impennis*).

throughout the Polar Regions. The appearance of the King Eider (*S. spectabilis*) on our coasts is an extremely rare occurrence, and even in Norway it is only known as a visitor, but on Novaya Zemlya and along the Arctic shores of Siberia, in Greenland and

Arctic North America, it is known to breed. The
third species, Steller's Eider (*S. Stelleri*), seems to be
still rarer, and only in the Aleutian islands and in the
north of Alaska can it be said to be at all abundant.
It is probable that the famous Great Auk (*Alca
impennis*, Fig. 9) also was a typical Arctic species.
Its range extended to both sides of the Atlantic. In
Newfoundland and on the coast of Iceland it is known
to have been met with in considerable numbers
within historic times; and no doubt, like all Arctic
species, it extended farther southwards at a more
remote period.

The members of the genus *Lagopus*, including the
various species of Grouse, are likewise of northern
origin. The British Red Grouse (*L. scoticus*), which
may be looked upon as a form of the Scandinavian
Willow Grouse (*L. albus*) (compare p. 91), constitutes
in some respects a curious case of parallelism with
the Arctic Hare, since the latter, in its more southern
station, generally retains the summer fur throughout
the year. The allied Ptarmigan (*L. mutus*) inhabits
Scandinavia, the Ural Mountains, and some of the
Asiatic mountain ranges. It is also found in the
European Alps and in the Pyrenees. The North
European range of the Ptarmigan suggests that
we are dealing with an ancient species which came
south from the Arctic Regions at about the same
time as the Arctic Hare; but it is more probable,
as I have shown in a subsequent chapter (p. 334),
that this species has entered Europe more recently

with the Siberian migrants from Central Asia, where indeed the genus had its original home. The Black Cock (*Tetrao tetrix*) and the Capercaillie (*Tetrao urogallus*) have also come to us from the east, and have even penetrated into Ireland. They are therefore some of the few instances of members of the Siberian invasion having become temporarily established there.

Reptiles and amphibia are altogether unknown in the Polar Regions, but a large number of fish, chiefly marine, have taken their origin there. The Salmon family are of Arctic origin, as also are the Stickle-backs and the Perches, many of the Cod family, the Herrings, and several of the Flat fish.

It would lead me too far to refer to the invertebrate fauna of the Polar Regions, but a few remarks on the Arctic plants may not be out of place.

The principal Arctic genera are Salix, Ranunculus, Draba, Pedicularis, Potentilla, Saxifraga, Carex, Juncus, Luzula, Eriophorum, and others.

Among the most characteristic Arctic plants may be mentioned *Dryas octopetala*, to which I have already referred as occurring in the west of Ireland; *Saxifraga oppositifolia*, another British species, occurs in the higher mountains of Scotland, Ireland, and Wales; *Braya alpina*, *Papaver nudicaule*, *Lychnis apetala*, *Diapensia lapponica*, and *Lobelia Dortmanna*, which is found in the lakes of Scotland and Ireland. The dwarf birch (*Betula nana*) also, which still occurs in Scotland and the North of England, and which had formerly a wider range in the British Islands, should be included among these; but there are other plants probably of Arctic origin, though not now occurring in the Arctic Regions, and to these may be classed the so-called American species of plants which are found on the northern and western coasts of Ireland, in the Hebrides, in Scotland, and in North America. These are no doubt the relics of an Arctic flora which flourished in high latitudes in past times when the climate there was more temperate A list of these species will be found on page 166.

As none of them occur in Siberia, they must either have found their way to North America and to Europe from the Arctic Regions, or have travelled from North America across the latter to Europe. In any case a former land-connection between the two continents must have existed. This becomes the more evident when we examine the remarkable results obtained by the late Professor Heer, who first described the Tertiary plant-beds in North Greenland. No less than 282 species of plants have been described by this eminent botanist from these deposits. A large number of the plants found were trees belonging to the genus *Sequoia*, *Thujopsis*, and *Salisburia*, besides beeches, oaks, planes, poplars, limes, and magnolias. That they grew on the spot is proved by the fruits, which have been obtained from these beds in various stages of growth.

From a similar deposit in Spitsbergen a large number of fossil plants have also been brought to light, many of which are identical with those found in Greenland; and some of the Greenland forms (such as *Taxodium distichum* and *Sequoia Langsdorfii*) have been found too in Alaska, showing that there was probably a continuity of land between Spitsbergen and North America by way of Greenland. Two species of *Sequoias*, namely, *S. sempervirens* and *S. gigantea*, the well-known Californian giant trees, are very closely allied to the Greenland forms discovered by Professor Heer.

Heer assigned the Arctic plant-bearing beds to the Miocene epoch, but doubts have been recently thrown upon this opinion by Mr. Starkie Gardner, who brought forward arguments in support of his theory of their being of the Eocene age. Professor Heer, however, was able to meet these criticisms, and he is ably supported in his views by Professor Engler and other eminent continental botanists.

It is evident that under the present conditions of temperature none of those plants could have flourished in Greenland. The climate must have been much milder than it is at present. Professor Heer estimated from the general aspect of the fossil flora that the mean annual temperature of North Greenland was at least nine degrees centigrade, and that the mean winter temperature was not below zero.

It will hardly be necessary for me to review here the various theories which have been advanced by geologists and botanists to account for this remarkably high temperature in such northern latitudes

10

Any one who has read the writings of the late Dr. Croll cannot help being struck by the facts he adduces to show the importance of ocean currents in relation to the distribution of heat over the globe, and it seems to me that the view which attributes the mild climate prevailing in former times in Greenland to warm ocean currents reaching the Polar Circle is the one least open to serious objections. If we suppose that the North Atlantic Ocean was bridged by a land-connection between Scandinavia and Greenland by way of Spitsbergen, and between Greenland and North America, the Polar Ocean would be practically a closed sea. If, then, a wide passage existed somewhere about Behring Straits to allow a warm current to enter and circulate within the Arctic Seas, we should have the southern shores of Greenland washed by the warm Atlantic current and the northern shores by a warm Pacific current, which combination would undoubtedly produce the effect of raising the temperature throughout the Polar Regions very considerably; and especially would that be the case with regard to Greenland and the neighbouring islands.

It might be urged that the constant darkness during winter must have had an injurious action upon the flora, but it is found that in countries such as Northern Russia, where southern plants are housed during winter in greenhouses, the light being almost entirely excluded by a covering of straw, no serious damage is done thereby to the plants.

It seems probable that a similar gradual refrigeration of climate in northern latitudes has taken place after Miocene times as has been proved to have occurred in Europe.

Some years ago Dr. Haacke propounded the hypothesis that the centre of creation of all the larger groups of animals was situated in the region of the North Pole, and that the newly originated groups must always push the older ones farther and farther south into the most remote corners of the earth. As instances of the correctness of his view he quotes the fact that the more ancient mammals, such as Monotremes, Marsupials, Lemurs, Edentates, and Insectivores, all inhabit the more southerly parts of the world. The Apteryx, Moa, Rhea, and the Ostrich, as well as Æpyornis, which is only recently extinct, are found in the same regions. But we have no palæontological evidence in favour of these extravagant views. Fossil Edentates and Marsupials are almost entirely confined to the Southern Hemisphere, and the supposition that because these primitive mammals inhabit the extreme south of our great continental land-masses, they therefore came from the north, cannot be said to be an argument. Nevertheless, I am quite with Dr. Haacke in considering that the North Pole, or, we might say, the lands within the Arctic Circle, have been the place of origin of some of our European mammals, and there can be no doubt that certain species in other groups, among invertebrates and also plants, have originated

in the Polar Regions. The facts of geographical
distribution teach us that in these regions there has
been a centre of origin within comparatively recent
geological times. I have on a previous occásion
drawn attention to the range of the Reindeer: that
it lives almost throughout the Polar lands, and that it
spreads into North America, Northern Europe, and
Northern Asia. We have, again, fossil proof that its
range extended down to the Pyrenees in Europe in
pleistocene times. But there is not a scrap of
evidence that it ever during any time occurred
farther south, either in Europe, Asia, or North
America. Its original home must therefore have
been in the Polar Regions, for if it had originated
either in Central Europe, Asia, or America, there is no
reason why it should not, in the natural course of
events, have extended its range to the south as well
as to the north.

The Arctic Hare presents us with a very similar
case of distribution. Like the Reindeer, it inhabits, as
we have learned, the Polar Regions and the northerly
parts of the Old World and the New; but while we
have only fossil evidence of the former, more southerly,
extension of the range of the Reindeer, the Arctic Hare
furnishes us with a still stronger proof of its past
southward range in the survival of small isolated
colonies in some of the southern mountain ranges of
Europe and Asia. It is generally believed that the
occurrence of the Arctic Hare in these southern mountains is a standing testimony to the severity of the

climate at the time when it commenced its southerly increase of range, but I have already shown that the climate of Europe at that time was not necessarily colder than it is at present, but that it may have been somewhat milder (p. 80). I think that a vast increase of ice in the Polar Regions has taken place only at a comparatively recent date, and that both the Reindeer and the Arctic Hare originated there during a much more temperate climate than obtains at present. A great sensation was produced among European zoologists and anthropologists when the discovery was first announced that the remains of the Reindeer had been found in the Pyrenees, and it naturally gave rise to many speculations as to the nature of the climate at the time when its range extended so far south.[1] The greater number of our best authorities are still of opinion that the existence of the Reindeer in Southern Europe points to the prevalence of an arctic climate in that region. It is generally overlooked, however, that the Reindeer-remains occur in company with many typically southern animals, which,

---

[1] A very interesting piece of information has been given us, recently, by Mr. Barrett-Hamilton on the Arctic Fox of Spitsbergen. In comparing the skulls of Spitsbergen Foxes with those of Europe, he found that the former are much smaller, and represent a distinct race or subspecies. This small race he believes to be confined to Greenland, Iceland, Spitsbergen, and Novaya Zemlya, whilst the larger one occurs in Europe, Asia, and on the Commander Islands. This fact favours the view which I have advocated in Chapter V., that the Arctic Fox in Europe is a Siberian migrant, and did not come from the north with

if they had been found alone, would have been held to be a certain indication of a warm climate. The French geologist Professor Lartet, indeed, was of opinion that the temperature during the time when the Reindeer lived in the Pyrenees must have been rather milder than it is at present (compare pp. 71-75). Similarly, Mr. Harlé argues, that the extremely cold climate probably did not extend to South-western France, since that area only received occasional visits from some of the representatives of the Arctic fauna.

Long ago North American zoologists recognised the existence in their country of two well-marked races of the Reindeer (Caribou)—a smaller one with rounded antlers (Fig. 10), and a larger one in which the antlers are more or less flattened out (Fig. 11). Two somewhat similar races can also be traced in the fossil remains of the Reindeer in Europe. It was, I think, Gervais who first pointed out that the Reindeer remains from the north of France differed from those found in the south; and Lartet referred to the fact that the southern remains were more like what, in America, is called the Barren-ground Caribou, while those from Central European deposits all belonged to the Siberian variety, which is more like the Woodland Caribou of North America. In Ireland, Professor Leith Adams also drew attention to the curious fact that all the Irish Reindeer remains resemble the Norwegian variety rather than the Siberian; and Mr. Murray was so much struck by

'IG. 10.—Head of a Barren-ground Reindeer in the Dublin Museum (photographed by Mr. McGoogan).

Fig. 11.—Head of a Woodland Reindeer in the Dublin Museum
(photographed by Mr. McGoogan).

the close resemblance between the Spitsbergen and Greenland forms with the Barren-ground Caribou, that he•based some speculations on a former land-connection between these countries on this circumstance.

We have, therefore, records of the present or the former existence of a Reindeer resembling the North American Barren-ground form in Greenland, Spitsbergen, Scandinavia, Ireland, and the South of France. In England the remains of the two forms occur mixed, but I do not know in how far either the one or the other predominates. The Barren-ground Reindeer is in Europe altogether confined to the west; the most easterly locality that I am acquainted with being Rixdorf, near Berlin. The majority of the European remains of the Reindeer seem to belong to the Siberian or Woodland variety, and it would appear as if some intercrossing between the two forms had occurred in Lapland, since it is stated that in that country the Reindeer is somewhat intermediate between the two. All the Asiatic remains also resemble the Woodland variety.

As far as I know, no explanation has been attempted to account for this peculiar range in Europe of the two forms of Reindeer. But if we look more closely into the mode of occurrence of the Reindeer remains, we find that the Barren-ground form, seems to have existed in Western Europe long before the other variety made its appearance there. It was pointed out by Struckmann that the Reindeer in Southern Europe occurs in older deposits than in the north. In speaking of the northern ones, he had of course chiefly the German deposits in view. It is in one of the oldest pleistocene deposits in Germany that the isolated instance, referred to above, of the occurrence of the Barren-ground Reindeer, near Berlin, has been noted.

There is still a further point which illustrates the supposition that the Barren-ground Reindeer was a more ancient inhabitant of Europe than the Woodland one. The latter in all Central European stations (in fact almost wherever it occurs fossil) is associated with the remains of the typical inhabitants of Siberia, such as the Glutton, Sousliks, Lemmings, and others; but in the deposits in which the Barren-ground Reindeer have been found in South-western France, no other Arctic mammal finds a place. Again, in Irish deposits none of the Siberian migrants are found. The only explanation of this remarkable fact is that the two varieties of the Reindeer have come to Europe by different routes. We have learned already from the observations of Mr. Murray that there are evidences of the existence of a former land-connection between North America, Greenland, and Spitsbergen. Professor Petersen tells us that, according to recent surveys, a high submarine plateau with a sharp fall of 1000 fathoms towards the Atlantic Ocean begins from Northern Norway and is continued as far as Spitsbergen. Several islands, such as Bear Island, King Charles Land, and others, arise from this plateau, and these must be looked upon as the remains of a sunken land (Fig. 12).

From Arctic America, thinks Professor Schulz (p. 1), we probably have had an uninterrupted migration during the greater part of later Tertiary times up to the commencement of the Pliocene epoch —partly over a direct land-connection between Greenland, Iceland, and the Faroes, and also between Arctic America, Spitsbergen, Franz Josef Land, etc. There was also a connection between Asia and Alaska.

The distribution of the Barren-ground Reindeer in Europe seems to warrant the belief that, at the time it began its southward wanderings from the Polar area, Northern Norway must have been connected with Greenland in the manner just indicated, but, as I shall explain later on, Russian Lapland and part of Northern Russia, or the land between the White Sea and the Baltic, must at that time have been submerged by the sea. The greater part of Denmark and the lowlands of Sweden were likewise submerged, but Scandinavia extended south as far as Scotland, while Scotland was connected with Ireland, and the latter with England and France. The Reindeer migrating south into Scandinavia could only reach the continent of Europe by way of the British Islands. It appeared there in the west and gradually extended its range east, where, as I mentioned above, it has occurred in a few isolated localities.

The advent of the Woodland form of the Reindeer in Europe took place at a much later stage. It came,

as I indicated, with the hordes of Siberian migrants
which invaded Europe during what is known as the
Inter-glacial phase of the Glacial period.   Scan-

FIG. 12.—Map of Europe, indicating the parts which were probably
submerged (shaded) at the commencement of the Glacial period.
The light portions represent, approximately, the extent of the
land at that time.

dinavia, not being then directly connected with continental Europe, was not accessible to it; neither
was Ireland, which had by that time become dis-

connected from Great Britain. None of the Siberian migrants seem to have been able to cross the River Garonne, and we therefore find neither the Woodland Reindéer nor any of the typical Siberian species represented in the Pyrenean deposits. The Woodland Reindeer persisted in continental Europe until comparatively recent times, and it has since made its way into Scandinavia across Northern Russia, and probably mingled with the older stock of the Barren-ground form. In the same way, it may have come about that in the English pleistocene deposits the remains of the two races occur.

In a recent contribution to our knowledge of the deer tribe (c, p. 88), Mr. Lydekker suggests that the former division of the Reindeer races into the two forms of Woodland and Barren-ground Caribou, no longer holds good. He now recognises no less than six races, as follows :—

1. Rangifer tarandus typicus.
2.   ,,   ,,   spitzbergensis.
3.   ,,   ,,   caribou.
4.   ,,   ,,   terræ-novæ.
5.   ,,   ,,   grœnlandicus.
6.   ,,   ,,   arcticus.

I hardly think these can be considered of equal value; indeed, though there may be differences between *R. grœnlandicus, typicus, arcticus,* and *spitzbergensis,* the antlers exhibit a certain much closer relationship among one another than to *R. terræ-novæ* and *caribou.* But the whole subject is by no means as well known as could be wished, and a very careful comparative study of recent and fossil remains of the Reindeer from various parts of the Old and New Worlds is much needed to put our views on a firmer basis.

The presence of the Arctic Hare in Ireland and the absence of the common European Hare (*Lepus europæus*) can be explained in a somewhat similar manner. The Arctic Hare is the older of the two species—corresponding with the Barren-ground Reindeer—and the European Hare the newer one, associating, like the Woodland Reindeer, in its westward migration with Siberian animals, though probably of Oriental origin.

Let us once more refer back again to the map on page 137 indicating the geographical distribution of the Arctic Hare. Its discontinuous range and its isolated position in the Alps, Pyrenees, and the Japanese mountains, all tend to show that it is an ancient species. Moreover, its presence in Ireland in the plain as well as in the mountains, clearly points to the fact that, in the British Islands at any rate, the Arctic Hare was the first comer, and that subsequently the European Hare invaded these countries. It probably found Ireland then no longer accessible, having since become separated from England. Again and again do we find the statement repeated, that the presence of the Arctic Hare in Europe is a clear proof of the former prevalence in our continent of an Arctic climate. But if so, why should this Hare at present live and

thrive in Ireland, which has a particularly mild climate
in winter, and be absent from so many continental
stations where the temperature more resembles that
of its native home? If we suppose that the European
Hare migrated to Europe from the east, after the
Arctic Hare had become established in Western
Europe, and drove the latter into the mountains
or northward whenever the two came into contact,
we should have, it seems to me, a better explanation of the range presented by the two species.
I was formerly of opinion that the European Hare
had come with the Siberian animals from Siberia,
but it appears to me more likely now, that it reached
our continent with the Oriental migrants, and only
then joined the Siberians in Eastern Europe.

The evidence in favour of a former land-connection
between Scandinavia and Greenland, rests on many
other facts besides those already brought forward.
That some form of land-connection formerly existed
between Europe and Greenland is now indeed almost
universally accepted. That it was situated more to
the south between Scotland and Greenland is a supposition which has been actively supported by many
leading authorities, but it seems to me that if such a
land-bridge existed, it must have been in very early
Tertiary times, whilst the northern one, such as I
have indicated, may have originated later and persisted until a recent geological date.

The distribution of few groups of animals is now
better known than that of the larger butterflies and

moths (*Macro-lepidoptera*); even those of Siberia have
been fairly well investigated. The interesting facts obtainable from their distribution are therefore of special
value. No less than 243 species of *Lepidoptéra* are
mentioned by Möschler as being common to North
America and Europe. It is extremely probable that
a fair number of these have either migrated direct
from America to Europe or *vice versâ*, though many
may be of Asiatic origin, and have wandered east
and west from their original home. The following
twelve species are mentioned by Petersen (p. 38)
as occurring in Arctic Europe and also in Arctic
North America, but not in Asia:—*Colias nastes,
Colias hecla, Syrichthus centaureæ, Pachnobia carnea,
Plusia parilis, Anarta Richardsoni, Anarta Schön-
herri, Anarta lapponica, Anarta Zetterstedti, Cidaria
frigidaria, Cidaria polata, Eupithecia hyperboreata;*
and these, as he remarks, point to the possibility
of a former direct land-connection between Europe
and North America.

Mr. Petersen believes that the chief immigration
into the Arctic area of Europe is post-glacial and
took place from Siberia, since the majority of the
species are still to be found in that country at the
present day (p. 57). He also draws particular attention to a fact,—which I shall discuss more fully in the
next chapter,—namely, that the most characteristically
Arctic forms of Northern Europe, which also partly
occur in the Alps, are entirely absent from the
Caucasus.

Adopting the glacial views of some of our leading geologists, Petersen comes to the logical conclusion that Central Europe could not have possessed any butterflies during the height of the Glacial period, but since all evidences seem to point to the chief migration from Siberia having taken place after the Glacial period, he concludes that they must have survived the severe cold of that time in Central Asia. He leaves us, however, to imagine under what possible geographical conditions the climate in Europe could be too severe for a lepidopterous fauna, while at the same time Central Asia could maintain an abundant one.

In a suggestive note on the origin of European and North American Ants, Professor Emery states (p. 399) that a great number of North American ants are specifically identical with European ones; whilst Dr. Hamilton tells us (p. 89), as an instance, that specimens of the beetle *Loricera cærulescens* from Lake Superior and from Scotland do not seem to vary to the extent of a hair on the antennæ. He enumerates 487 species of *Coleoptera* as being common to North America, Northern Asia, and Europe, many of which no doubt have migrated by the Americo-European land-connection.

Arctic Scandinavia or Lapland, according to Sir Joseph Hooker, contains three-fourths of the entire number of species of plants known from the whole circumpolar area. His view, that the Greenland flora is almost exclusively Lapponian,—having only

11

an extremely slight admixture of American or
Asiatic types,—again points to a former more
intimate connection between North America and
Arctic Europe, and indeed he remarks (p. 252),
" It is inconceivable to me that so many Scandinavian plants should, under existing conditions
of sea, land, and temperature, have not only found
their way to Greenland by migration across the
Atlantic, but should have stopped short on its
western coast and not crossed to America."

Hooker's view, that the Scandinavian flora is of
great antiquity, that, at the advent of the Glacial
period, it was everywhere driven southwards, and
that during the succeeding warm epoch the surviving species returned north, has been adopted by
the great majority of naturalists.

The natural corollary of this theory is that there
must have been, between the beginning of the Glacial
period and the present time, either two independent
land-connections between the Polar Regions and
Northern Europe at different epochs to enable
animals and plants to travel southwards and once
more to regain their former northern home, or,
that during the whole of the Glacial period the
Polar Regions were uninterruptedly connected with
Northern Europe, until the fauna and flora had once
more reached their northern goal, after the Polar
lands had been desolated by the supposed rigours of
that period.

In following the history of the Arctic migration to

Europe, it is of great importance to determine the nature and the time of duration of these land-connections. The Greenland flora is a very instructive one in helping us to understand many of the problems connected with the origin of the European plants and animals. To judge from the remarks of Professor James Geikie and Mr. Clement Reid, no flowering plants could have existed in the British Islands during the height of the Glacial period, and one would suppose that the cold in Greenland at that time must have been far more intense than in England. If no flowering plants could exist in the latter country, then very surely none could in Greenland, where the climate was of necessity by far more rigorous. It will be a surprise, therefore, to those who are acquainted only with Professor Geikie's views of the nature of the Glacial period, that two of the most eminent Swedish botanists, who have made a special study of the flora of Greenland, have come to the conclusion that a survival of flowering plants has taken place in Greenland itself from pre-glacial times. According to Professor Nathorst (p. 200), only a few plants could have survived the Glacial period in Greenland. The species now peculiar to that country may perhaps, he thinks, be the remnants of those which existed in pre-glacial times. Mr. Warming, on the other hand, is of opinion that the main mass of Greenland's present flora survived the Glacial period there (p. 403), and that the remainder was carried from Europe and North America by

occasional means of distribution of the nature indicated by Darwin.

Very similar views on the origin of the present
Polar flora are expressed by Colonel Feilden, who
says, " To my mind it seems indisputable that several
plants now confined to the Polar area must have
originated there and have outlived the period of
greatest ice-development in that region" (b, p. 50).
No land-connection at all need be supposed to have
existed in recent geological times, that is to say,
during the Glacial period or after, if Mr. Warming's
and Colonel Feilden's views be adopted. A pre-glacial
connection would be sufficient to explain the general
features of distribution. An admission is thus obtained from these two independent authorities that
the climate during the Glacial period must have
been vastly less severe in the Polar Regions than
is generally conceded. I am of opinion that not
only the whole of the present flora, but also the
fauna of Greenland survived the Glacial period in
that country.

If we suppose that an extensive centre of origin
existed in the Polar area, or we may say in Greenland, both animals and plants would have been able
to spread from it into Northern Europe and North
America by means of the land-connections which are
generally supposed to have existed in pliocene times,
that is to say, just before the commencement of the
Glacial period. There must have been at this time
a connection too between Scotland and Scandinavia,

which will be dealt with more fully presently. The important point is to consider what light the Greenland flora and fauna will throw upon the problem of the continuity of the aforesaid land-connection during the Glacial period. We have seen that the Barren-ground Reindeer, a typically Polar species, penetrated as far south as the Pyrenees, the Arctic Hare went as far, while a number of other species of Polar animals and also of plants occur in the Alps. Of these it remains to be seen how many have come direct by way of Northern Europe or from the Polar Regions by way of Asia. At any rate, as the origin of the Alpine animals and plants will be discussed in another chapter, there is no need to dwell on this subject at present.

From the nature of the distribution in Ireland of Arctic plants and animals, which occur mostly on the north and west coasts, it would seem that a stream of migration entered from Scotland, and I have no doubt that that same migration came into Scotland directly from Scandinavia by a route over which now roll the waves of the North Sea. There is, moreover, as I already mentioned on p. 94, a very interesting so-called American element in the north-western European flora, that is to say, plants now found in North-west Europe and North America without occurring in Greenland or any of the islands which might have formed the former highway between the Old World and the New. These are probably some of the more ancient Polar plants which have become extinct in the Arctic Regions and survive in isolated patches in favourable localities. We find seven species of these American plants in Ireland, almost entirely confined to the north and west coasts. These are *Spiranthes Romanzoviana, Sisyrinchium anceps, Naias flexilis, Eriocaulon septangulare, Juncus tenuis,* and *Polygonum sagittifolium.* To them must be added another plant recently discovered by the Rev. Mr. Marshall in the south of Ireland, namely *Sisyrinchium californicum.* As I have mentioned in former writings, there are three species of North American freshwater-sponges in Ireland which have not hitherto been discovered elsewhere in Europe or in Asia. These, namely *Ephydatia crateriformis, Heteromeyenia Ryderi,* and *Tubella pennsylvanica,* all occur in some of the lakes near the western coast of Ireland.

There are in all groups of animals instances of species which are confined to Europe and North America, while unknown from the Asiatic continent, but none, as far as is known, have such a very discontinuous range as that of the animals and plants just referred to. In some cases the species still occur in Greenland, and in this way make it still clearer that their migration in former times took place from one continent to the other by way of that country. As an interesting instance of such distribution may be mentioned the Common Stickleback (*Gasterosteus aculeatus*), which is found in Greenland, North America, and Europe, but is quite absent from Asia. Then again, the Nine-spined Stickleback (*Gasterosteus pungitius*) is confined to Western Europe and North America, though an allied species, *Gasterosteus sinensis*, lives in China and has probably penetrated there from the New World across the old Behring Straits land-connection.

The Coleoptera *Diachila arctica*, *Elaphrus lapponicus*, and *Blethisa multipunctata* are good instances of species which have come to us from North America by way of Greenland. I have already referred to the Lepidoptera, but might add that eleven species of *Anarta* occur in Scandinavia, eight of which reappear again in Labrador, none of them, however, being met with in Siberia. Then again, take the interesting Crustacean *Lepidurus* (*Apus*) *glacialis*. It is found in Greenland, Spitsbergen, Lapland, and Norway; and formerly, as we know from fossil evidence, it ranged into Scotland. Another Phyllopod, viz., *Branchinecta paludosa*, inhabits Greenland, Lapland, and Norway. Mr. Kennard suggests that the freshwater Snail *Planorbis glaber* might also belong to the same migration. And there are no doubt large numbers of others.

Professor Emery mentions that Northern Europe possesses one peculiar genus of Ant, viz., *Anergates*. This is closely allied to *Epoccus*, another genus confined to North America. It seems probable, therefore, that both of these have sprung from an Arctic genus which sent two branches southward into the two continents without there being any migration through Asia.

The general range of the Arctic plants and animals

gives no reason to suppose that the Greenland fauna
and flora of the present day were exterminated by the
Glacial period and then reintroduced into that country.
Nor have we any evidence that such a fauna and flora
migrated across the British Islands northward. The
Greenland animals and plants too are altogether much
more like the Lapland ones than those of Scotland.
It will also become evident to the reader of this work
that no very extensive migrations could have taken
place during the post-glacial period, and that almost
everything points to a survival of both fauna and
flora in northern latitudes throughout the Glacial
period.

If we take into consideration the palæontological
evidence of the two races of Reindeer in Europe, one of
which came to us from the north, and that the Arctic
Hare and one of the races of the Stoat entered our
continent from the same direction—when we, moreover, carefully review the numerous other instances
quoted of plants and animals which could only have
reached us from the north, the irresistible conclusion is
forced upon us that a land-connection existed at no
very distant period between Northern Europe and the
Arctic Regions of North America. This is not a new
hypothesis. Many geologists are of opinion that a
land-passage did exist within comparatively recent
times, uniting Europe, Greenland, and North America.
But the position of this old land-bridge, as I have
mentioned, has been generally placed somewhat
farther south than I should feel inclined to put it.

The fact that very extensive glaciers formerly
covered the mountains of Scandinavia on the eastern
side, whilst they scarcely reached the sea on the west
(Feilden, *a*, p. 721), seems to favour the view of a
warm current having washed the western shores. As
I shall attempt to show later on (p. 179), the Arctic
Ocean extended across Northern Russia at that time
from the White Sea to the Baltic—that is to say, to
the eastern shores of Scandinavia, which country was
then joined to the north of Scotland. The predis-
posing agents to a copious snowfall existed in
Scandinavia, viz., an excessive evaporation of the
warm Atlantic waters and unusual precipitation in
the form of snow owing to the cold given off by
the Arctic waters on the east side of the mountains.
It is therefore probable that the land-connection
which united Europe and North America was farther
north than has been supposed.

If we sail straight across from Northern Scandinavia
to Greenland, we traverse an exceedingly deep marine
basin; but if we examine the sub-marine bank which
runs all along the coast of the former country from
south to north, we find that it does not end when the
extreme north of the land is reached. The bank
extends much farther north, and is continued as far
as Spitsbergen. As I have said before, the latter,
as well as Bear Island, must be looked upon as the
remains of a large mass of sunken land—the ancient
Scandinavia stretching far into the Arctic Circle. Professor Nathorst speaks of Spitsbergen as a northern

continuation of Europe, not only geographically, but also botanically and geologically.   However, this northern land must have stretched even farther—not

FIG. 13 —Map of Europe, indicating approximately the distribution of land and water during the earlier stages of the Glacial period— shortly after the period represented in Fig. 12, p. 156.   The darkly shaded parts indicate the areas covered by water, and the white portions what was land at the time.

perhaps farther north, but farther west.   Here lay the old land-connection between Scandinavia, Greenland, and North America (Fig. 13).   One of the highest authorities on the geographical distribution of plants, Professor Engler, maintains that the arguments in favour of this Arctic connection of America with Europe are more weighty than those for a land-bridge between Greenland, Iceland, the Faroes, and Great Britain. Moreover, he is of opinion that a certain number of species of plants belonging to the Alpine flora of Arctic Siberia have travelled from Scandinavia *via* Greenland and North America to Eastern Asia, and not direct from Scandinavia to Siberia (p. 143).

That this ancient Arctic land-connection existed almost throughout the Glacial period appears to me probable. It has often been suggested that such a land-barrier was one of the principal causes of the production of the glacial phenomena in Europe, and as such it must have existed intact certainly during the earlier stages of the Glacial period. The barrier must then have gradually subsided in one or two places; and once a breach was formed, the complete union between the Atlantic and the Arctic Oceans could not have been long delayed.

The terrestrial fauna and flora, as we have seen, lend strong support to the view of the former connection between Scandinavia and Greenland, but many other facts point in the same direction. It was Edward Forbes who first drew attention to the presence of a number of species of littoral molluscs on the coast of Finmark which also occur on the coast of Greenland, and he expressed the firm conviction that

they indicated by their existence on both sides of the
Atlantic some ancient continuity of the coast-line.
He held that the line of migration of these mollusca
was probably from west to east, and that it must
have taken place during physical conditions entirely
different from those prevailing at present. If Forbes's
view is correct, a current must have existed from the
north coast of North America along the northern
shore of the ancient land which stretched east as
far as Europe. We have also some palæontological
evidence bearing on the existence of such a current
(p. 173).

As we shall learn presently, the early stages of
the Glacial period were accompanied by a marine
transgression over Northern Russia and Germany—
an overflow, as it were, of the waters of the Arctic
Ocean covering a great part of Northern Europe,
with the exception of Norway. One continuous
ocean ultimately extended from the east coast
of England across Holland, Northern Germany, and
Russia to the White Sea (Fig. 12, p. 156). The
south of England being at that time joined to
France, and Scotland to Scandinavia, there was no
direct communication between this large North
European Sea and the Atlantic. The glaciers
which took their origin in the Scandinavian Mountains discharged icebergs into this sea, and many of
them no doubt were stranded on the east coast of
England. The boulders of Scandinavian origin which
have been discovered in recent geological deposits on

that coast have generally been traced to the action of land-ice, but the supposition that they have been carried by icebergs—the older theory—appears to me the more probable one. Such boulders begin to make their first appearance in the Red Crag, a deposit which is now looked upon as belonging to the newer pliocene series. But whether we call it pliocene or pleistocene really matters little. The important fact is, that glacial phenomena, consisting of the appearance of boulders foreign to the country together with an invasion of Arctic shells, are now ushered in upon a coast which shortly before teemed with the southern life of a Mediterranean character. Among the new arrivals in these English crags there are no less than eighteen species of North American marine mollusca. Since the German Ocean had then no direct communication with the Atlantic, these mollusca could only have come from the White Sea, and Forbes's *Arctic current* would offer an explanation of the manner in which they were enabled to migrate there from their original home.

It might be urged that we have no grounds for the supposition that the German Ocean was practically a closed basin; and that these American species probably inhabited at that time the whole of the North Atlantic Ocean. But if such had been the case, we ought to have evidence of the occurrence of some of these species in the newer Tertiary deposits along the west coasts of the British Islands. Such beds exist; there is, however, not a trace in any of them of any American mollusca. In examining the marine deposits of St. Erth, on the coast of Cornwall, which are believed to be of about the same age as the newer crags, Messrs. Kendall and Bell were much struck by the absence of the species characteristic of the latter. The St. Erth fauna led them to believe that the Arctic Ocean could not then have opened into the Atlantic, but that a land-communication had existed between Europe and North America, so as to form a barrier of separation between the two oceans. This again perfectly harmonises with the views I have expressed, and supports them.

Let us now look a little more closely at the history and the fauna of the Baltic and the adjoining lakes, in order to gain additional information as to the geographical changes which have had such lasting influence on the peninsula of Scandinavia. The Baltic is a shallow sea covering an area of 184,496 square miles, and its waters are decidedly brackish. The fauna is a poor one, being too salt for the purely freshwater species and not salt enough for the typical marine forms. The absence of some animals which we should expect to find there is one of the remarkable features about the Baltic, but, on the other hand, some species occur which are altogether strangers to the fauna. And these, moreover, are confined to the extreme northern end of the sea. I need only refer to the Arctic Seal (*Phoca annelata*), which is confined to the Gulf of Bothnia, and to the four-horned sting-fish (*Cottus quadricornis*, Fig. 14, p. 178), neither of which occur on the west coast of Scandinavia. But there are others which point in an equally unmistakable manner to the former existence of a marine connection between the Baltic and the southward prolongation of the Arctic Ocean—known as the White Sea. It is generally admitted now that such a union between these two seas, viz., the Baltic and the White Sea, occurred in recent geological times, but opinions differ as to the duration of this connection. I adhere to the view expressed by Murchison and others, that the boulder-clay is a marine deposit. I am also convinced that the Arctic Ocean, as I have already mentioned, transgressed over the lowlands of Northern Russia at about the time when the newer crags were being deposited on the east coast of England; that the same large sea also covered Northern Germany, Denmark, Holland, and the low-lands of Sweden, and laid down the lower continental boulder-clay which is spread over such vast tracts of land in those countries. I shall have occasion to refer to this again more fully in the next chapter; mean-while, it should be remembered that this stage was followed by a partial retreat of the northern sea, though Scandinavia did not become joined to the Continent. The date of this retreat of the sea, repre-sented in Fig. 13, corresponds probably to what is know as the inter-glacial phase of the Glacial period, and I think it must have been during this time that the Forest-Bed on the coast of Norfolk was laid down.[1]

None of the Siberian mammals apparently entered Scandinavia at the time when they invaded Central Europe and penetrated as far west as England and Western France. Nor did the great Oriental mammals, like the Mammoth and others, reach Scandinavia; and Professor Pohlig argued, on the strength of these facts, that the latter country was either for a very short time only free from ice, or that it had defective land-communication with the Continent during inter-glacial times. This seems to me scarcely to explain the facts of distribution and account satisfactorily for the absentees. Nor does it, of course, harmonise with the views that I have announced above. Professor Engler's remark (p. 131), that Scandinavia probably projected above the glacial sea as an island, is more in accordance with these views, though the term island is scarcely applicable to that country, since it was always, as I said, indirectly joined to the Continent (*vide* Fig. 13, p. 170). The fauna of Scandinavia, both fossil and recent, points to a direct isolation of that country from the continent of Europe during a considerable period.

Another proof that Northern Russia and the lowlands of Sweden were covered by the sea comes to us from a study of the fauna of the relict lakes—the "Reliktenseen" of Leuckart. This name was first applied by Leuckart to lakes containing marine

---

[1] I have already expressed this view on p. 120.

organisms, which are supposed to have been flooded by, or to have been in close communication with the sea at some former period, like the lakes Ladoga and Onega in Russia. His views have been worked out subsequently in greater detail by Lovén and O. Peschel, who gave them their strong adherence. Many leading zoologists, such as Professor Sars and others, have since adopted them, and though discredited by Professor Credner, the theory still offers the best explanation for the origin of marine animals in freshwater lakes.

Professor Credner's contention, that marine mollusca are always absent from these relict lakes, seems at first sight a stumbling-block to the theory. But the explanation is really simple enough. It is to Dr. Sollas that we owe a very ingenious explanation of the origin of freshwater faunas. He showed that all freshwater organisms in their early stages of development are provided either with some process enabling them to attach themselves to a foreign object, or that they pass this period within the body of the parent. This is a provision of nature to prevent freshwater organisms from being floated out to sea, where they would perish, until they reach maturity and can cope with floods and currents. Had Professor Credner been aware of Dr. Sollas's views, no doubt he would have modified his criticisms, for, as most marine mollusca have free-swimming larvæ, they would have little chance of becoming permanent residents of lakes. During their larval stage, marine molluscs are quite a

prey to the currents of the sea. They have practically
no swimming organs, and only move by lashing to and
fro the tender cilia with which they are provided. ·

This disposes, therefore, of Professor Credner's
main criticisms. As for the fauna of the relict
lakes, we are now only concerned with those of
Northern Russia, Finland, and Sweden. In the
lakes Wetter and Wener in the latter country
occurs the four-horned sting-fish (*Cottus quadricornis*,

Fig. 14.—The Four-horned Sting-fish (*Cottus quadricornis*), reduced
from Professor Smitt's figure in the *Fishes of Scandinavia*.

Fig. 14), which, as we have learned, also inhabits the
northern part of the Baltic, and, as was suggested,
migrated there at a time when the latter was
connected with the White Sea. The principal food
of this little fish consists in a marine Crustacean
called *Idotea entomon*, an animal allied to our
common woodlouse. This is a typical marine
species, but it occurs also in the relict lakes of
the countries mentioned above, as well as in the
Baltic and the Caspian. Perhaps the best known

rm with a similar range is the Schizopod crus-
cean *Mysis relicta*[1] (Fig. 15), which is clearly a des-
ndant of the Arctic marine *Mysis oculata*, of which
was formerly considered a mere variety. The two
mphipods *Gammaracanthus relictus* and *Pontoporeia
finis*, and the Copepod *Limnocalanus macrurus*, are
ree additional well-known Arctic crustaceans whose
nge differs but little from those above-mentioned.[2]

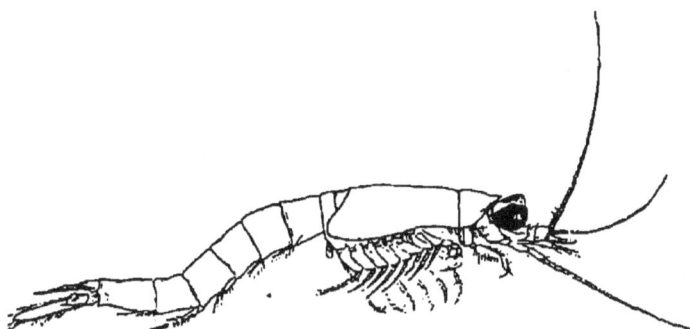

G. 15.—*Mysis relicta*, a small shrimp-like Crustacean, after Sars
(enlarged).

These facts all go to prove that the sea formerly
overed the lowlands of Sweden, Finland, and
orthern Russia. The fauna of Scandinavia, as
e have seen, indicates that during the greater part
f the Glacial period the country was not directly
onnected with continental Europe as it is now. It
eems that the barrier of separation probably con-

[1] The occurrence of this species in Lough Neagh in Ireland, pointing
a connection between the Irish Sea and the Baltic, will be referred to
ter on; as also that of two allied forms in the Caspian Sea.
[2] For additional species with a similar range, *vide* Nordquist.

sisted of a broad expanse of ocean on which floated numerous icebergs, which originated from the Scandinavian glaciers as they reached the sea. This was a cold sea, whilst Western Scandinavia was washed by the Gulf Stream (*vide* Fig. 12, p. 156). We might look upon the boulder-clay which covers such vast tracts of country in Northern Germany, Russia, and Holland as deposits formed by this sea rather than the ground-moraine of a huge Scandinavian glacier. I shall refer to this subject again in the next chapter; meanwhile it may be remembered that the boulder-clay of Northern Europe exactly resembles in all important particulars the similar accumulations met with in the British Islands. They resemble one another also in the occasional occurrence of sea-shells, the frequent appearance of bedded deposits, and the often inexplicable course taken by boulders from their source of origin. There occurs often a singular mixture and an apparent crossing of the paths of boulders in the boulder-clay. Professor Bonney remarks (p. 280) that these are less difficult to explain on the hypothesis of distribution by floating ice than on that of transport by land-ice, because, in the former case, though the drift of winds and currents would be generally in one direction, both might be varied at particular seasons. So far as concerns the distribution and thickness of the glacial deposits, he says there is not much to choose between either hypothesis; but on that of land-ice it is extremely difficult to explain the

itercalation of perfectly stratified sands and gravels
nd of boulder-clay, as well as the not infrequent
gns. of bedding in the latter. Two divisions are
enerally recognisable in the continental boulder-clay
-a lower and an upper. An inter-glacial phase
naracterised by a less severe climate is assumed
) have intervened between the deposition of the
vo. In Russia no such division can as a rule be
nade out, and sea-shells are either entirely absent or
xtremely scarce. It has been pointed out by Pro-
essor J. Geikie that the erratics—a name applied to
oulders in boulder-clay—in the upper division have
:avelled in a different direction from those contained
ı the lower. Taking for granted that the boulder-
lay is a marine deposit, this phenomenon seems
) indicate that the current which prevailed during
ıe early part of the Glacial period in this North
luropean ocean was different from the prevailing
urrent during the latter part. I have attempted
) explain this circumstance by the supposition
hat during the early part of the Glacial period
he Northern Sea had a connection with the Ponto-
Caspian Sea—a sea formed by the junction of the
3lack Sea and the Caspian (Fig. 12, p. 156). There
s geological evidence, as will be explained in the
ollowing chapter, that the area of these two seas was
onsiderably larger in glacial times than it is now,
.nd that they were joined across the valley of the
Manytch. After the inter-glacial phase of the Glacial
)eriod, the North European Ocean became connected

with the Atlantic Ocean across the north of England (Fig. 6, p. 126), the junction between the former and the Ponto-Caspian having meanwhile become dry land (Fig. 13, p. 170). A fresh current, now flowing westward, was set up in the North European Ocean, which accounts for the fact just cited that the erratics in the upper continental boulder-clay have travelled in a different direction from those in the lower. The boulder-clay laid down by the sea on the midland and northern counties of England, just as was the case with the similar deposit on the Continent, is generally accredited to the action of land-ice. It is by most geologists looked upon as the ground-moraine, partly of the huge Scandinavian glacier which is supposed to have impinged upon the English coast, partly of local British glaciers.

But renewed geological investigations on this point throw doubts upon these theories. Thus Mr. Harmer remarks in a recent contribution to glacial literature (p. 775), that "it is difficult to see how the Baltic glacier could have reached East Anglia, though ice-floes with Scandinavian boulders might easily have done so, while had the Norwegian ice filled the North Sea and overflowed the county of Norfolk, some evidence of its presence ought to be found in the glacial beds of Holland."

All the phenomena of distribution of the British fauna and flora are, as we have seen, much more easily explained by the supposition of a damp, temperate

mate, such as might have been produced by the
oximity of a cold sea on one side and of a warm
e at the other, than by invoking an arctic climate
th enormous glaciers. Most of the living animals
d plants would have been exterminated under the
ter conditions. Palæontological evidence in Great
itain clearly indicates that southern species migrated
st to these islands, that Arctic species were then
iven south from their native lands,—probably owing
insufficient food-supply and climatic changes in
e north,—that finally eastern species invaded the
untry—all this without the annual temperature
Europe being apparently much affected. For we
id in the British pleistocene deposits—and Mr.
ydekker draws particular attention to this remark-
le fact—a curious intermingling of southern and
rthern mammals, which undoubtedly lived side by
de. Everybody knows that northern and Arctic
ecies can live perfectly well in a temperate climate,
it that it is almost impossible to acclimatise
uthern animals in an Arctic or even temperate
ie. We have in this circumstance almost a proof,
terefore, that the climate cannot have been very cold.
hough a cold sea bathed the shores of Eastern
ngland, and even eventually invaded a portion of
Iorthern England, the warm ocean on the west
iust have effectually prevented any great lowering ·
f temperature.

At the time when the North European Sea flooded
portion of England, Scandinavia was still connected

with Scotland, and the latter with Ireland (Fig. 6, p. 126). There is no doubt that the food-supply in the Arctic Regions was decreasing with an increase of snowfall and with the gradual lowering of the land, which reduced also the habitable area. Arctic species therefore were driven south in search of fresh pastures. But it need not be supposed that anything like a vast destruction of the fauna of the Arctic Regions took place. Only fewer mammals were able to find food in a given space than heretofore. This southward migration may have commenced, in the case of plants and the invertebrates, at a much earlier time, —during the Miocene or Pliocene Epochs,—but it is doubtful whether the mammals and birds which we find in our pleistocene and recent deposits began to travel south much before the commencement of the Glacial period. The beginning of the Glacial period in England, I think, is indicated by the deposition of the Red Crag, though the latter is generally regarded as belonging to the pliocene series. Much of the northward migration from the British Islands of Lusitanian and other forms had then ceased, but we have in Scandinavia, just as in these islands, a southern relict fauna and flora, plants and animals which had wandered across what is now the German Ocean from Scotland to Scandinavia, and have never become extinct in that country to the present day. I need only mention the Red Deer, the Badger, and Slugs of the genus *Arion*.

Professor Blytt directs attention to some such
uthern relict species of plants now only found in the
:treme south-west of Scandinavia, such as *Asplenium*
arinum, Hymenophyllum Wilsoni, Carex binervis,
:illa verna, Erica cinerea, Conopodium denudatum,
'eum athamanticum, and *Rosa involuta* (p. 28).
The Arctic fauna and flora in Scandinavia—that
to say, the descendants of those species which
igrated direct from Greenland and Spitsbergen,
. we have seen, are numerous. They of course per-
sted throughout the Glacial period in the country,
id are now in many localities being exterminated
artly by change of climate, partly by a keen com-
:tition with more vigorous rivals which have come
i Scandinavia from the east. It is a curious circum-
ance, as pointed out by Professor Blytt, that the
.rctic plants in the Botanic Gardens at Christiania
:e able to stand almost any amount of sunshine,
ut are very liable to be injured by the frost, and have
i be covered in the winter. A similar observation
as been made in the case of the Alpine plants at
:ew Gardens, which have to be wintered in frames,
ough their homes are either in the high Alps—among
ie everlasting snows—or in the intensely cold climate
f Greenland. Many of the Scandinavian plants ex-
ibit instances of discontinuous distribution, thus show-
ig their ancient origin; and there is altogether nothing
i the fauna and flora of that country which might
:ad us to believe that these were exterminated
uring the Glacial period and reintroduced subse-

quently. The climate during that period in Scandinavia was probably more equable and moister,— with a greater snowfall in winter and with less sun to melt the snow during summer,—so that the development of glaciers took more formidable dimensions, chiefly on the east side. The lowlands of Sweden were covered by the sea, whilst many of the valleys were choked with great glaciers, which cast off portions of ice as they reached the sea, just as the Greenland and other northern glaciers do (*vide* p. 237). A country which at the present day probably somewhat resembles the former Scandinavia climatically is Tierra del Fuego, in the extreme south of South America. Though there is an abundant snowfall, so that glaciers reach the sea in many parts of the country, the flora has been described by travellers as luxuriant; and it appears that the fauna also is richer than might be expected from the cheerless climate.

Towards the latter part of the Glacial period the land-connection between Scandinavia, Spitsbergen, and Greenland broke down, and the waters of the Arctic and Atlantic Oceans joined. Whether it was at this time or later that the other land-connection between Scandinavia and Scotland collapsed is difficult to determine; but it is certain, I think, that Scotland was still united with Ireland even after these two great land-bridges ceased to exist.

SUMMARY OF CHAPTER IV.

he fauna of the Arctic Regions is much poorer than that of the
er regions which are dealt with in this work. In some groups,
h as Reptiles and Amphibia, there are no representatives at all,
no doubt a larger number of species existed there in earlier
tiary times. At least we have fossil evidence that during
Miocene Epoch plants of many families flourished in Green-
d of which no vestige is now left in the Polar area. Climatic
ditions must therefore have changed, as in Europe. A
dual refrigeration took place, owing probably to the slow
hdrawal of the current which supplied the Arctic Sea with
mth. Greenland and Europe were then connected, and
Arctic Ocean was separated from the Atlantic. This land-
nection is supposed to have lain far north between Scan-
avia, Spitsbergen, and Greenland, and must have persisted
il towards the end of the Glacial period.

ks the temperature decreased and the land-area available in
north diminished, the surplus population, consisting of
mals and plants, and possibly also of human beings, moved
thward. We have traces in Europe, and especially in the
tish Islands, of a very early migration from the north in the
called American plants and in the freshwater sponges. The
graphical distribution of some of the Arctic species of mam-
ls is referred to in greater detail, to show how the relative
: of their entry into Europe can be determined. Two forms
Reindeer, resembling the Barren-ground and Woodland
icties, have been met with in European deposits, but only
: former occurs in Ireland and the south of France, whilst
tward the other becomes more common, and finally is the
y one found. It is believed that the Barren-ground is the
er form as far as Europe is concerned, and that it came to
with the Arctic migration, and that the other Reindeer
iched Europe much later from Siberia, when Ireland had
eady become detached from England. The range of the

Arctic Hare is equally instructive. It must have been a native of Europe since early glacial or pre-glacial times—before the common English Hare had made its appearance in Central Europe. Along with other Arctic forms, it entered Northern Europe directly from the Arctic Regions, by means of the former land-connection which joined, as I remarked, Lapland with Spitsbergen, Greenland, and North America. There need not have been a post-glacial connection between Europe and Greenland; the present flora of that country may have survived the Glacial period in the Arctic Regions, as has been maintained by some botanists and other authorities. Professor Forbes argued from the occurrence of the same species of shore mollusca on the coast of Finmark and Greenland that these two countries were not long ago joined, so that a slow migration from west to east along an ancient coast-line could have taken place. That such a migration actually occurred is further made probable, judging from the presence of American mollusca in the Crag deposits on the east coast of England. These came into the North Sea in the first place direct from the Arctic Ocean at a time when the two oceans freely communicated with one another across the lowlands of Northern Russia, Northern Germany, and Holland. Arctic shells are also found below the boulder-clay on the Baltic coast, and a free communication such as indicated is generally held to have taken place at no very distant date. The so-called "relict species" —marine animals left in freshwater lakes in districts formerly covered by this sea—lend some support to this view. But the view that the continental boulder-clay is a marine deposit is not now held except by a few, though I here bring it forward again, as it seems to me to fit in so much better with the known facts of distribution. The sea just referred to probably existed throughout the greater part of the Glacial period; and icebergs, which originated from the Scandinavian glaciers, would have brought detritus and boulders to the lowlands. Scandinavia was then connected with Scotland, and England with France.

CHAPTER V.

THE SIBERIAN MIGRATION.

In dealing with the British fauna in particular, I have drawn attention to the fact that it is chiefly in the south of England that we find fossil remains of eastern species of mammals in recent geological deposits. We can actually trace the remains of these species and their course of migration across part of the Continent towards Eastern Europe, and as none of their bones have been discovered in the southern or northern parts of our Continent, it must be assumed that their home lay in Siberia, where many still exist to the present day, and where closely allied forms also are found. Some of these Siberian migrants have remained in England and on the Continent to the present day. Many have become extinct. But the animals forming this eastern migration did not all originate in Siberia, though I have sometimes spoken of them collectively as Siberian migrants. There must have been other centres of dispersion of species in Europe. We know that a very active centre of development—at any rate for land-mollusca—lay in South-eastern Europe, either in the Caucasus or in the Balkan peninsula, or more

probably in both. The Alps no doubt produced a
number of species which have spread north and
south, and may in their wanderings have joined 'the
Siberian migrants in their western course, and thus
have reached the British Islands. Nevertheless, the
majority of the mammals belonging to the eastern
element of the British fauna (*vide* p. 95) have undoubtedly originated in Siberia. The Polecat (*Mustela putorius*) and the Harvest Mouse (*Mus minutus*),
for instance, are members of that eastern migration.
Both occur throughout Central Europe and a large
portion of Siberia, but are absent from the extreme
north and south of Europe and also from all the
Mediterranean Islands. A Siberian species, which
has never penetrated so far west as the British
Islands, nor even so far north as Scandinavia or
south to Italy, is what is known in Germany as the
"Hamster" (*Cricetus frumentarius*), a little Rodent
which spends the winter asleep in its burrows, and
surrounds itself with a great accumulation of food-
material carried there during autumn. The common
English Hare, which I formerly regarded as an
instance of a Siberian mammal, must now find a
place among the Oriental migrants. Its history is
very instructive, and I shall have an opportunity later
on to refer to it again. Meanwhile, it may be mentioned that though this Hare inhabits Europe in two
varieties or races, one of which, *Lepus mediterra-
neus*, is confined to Southern Europe, the latter owes
its origin to an earlier migration from Asia.

When we come to consider the eastern birds, we
have to distinguish between resident species and
migratory ones. The Black-throated Thrush (*Turdus
atrigularis*), which has been twice obtained in the
British Islands, is a mere straggler to Europe, and
is not known to breed there at all. Better known
birds, perhaps, are the Golden Thrush (*Turdus varius*),
which has even occurred as far west as Ireland, the
Rock-Thrush (*Monticola saxatilis*) and the Scarlet
Grosbeak (*Carpodacus erythrinus*), which breed in
Eastern Europe, but are known only as occasional
visitors in the west.

To judge by their distribution, the Bullfinches
(*Pyrrhula*) are of Asiatic origin, for seven species out
of ten are confined to that continent. Our common
Bullfinch (*P. europea*) probably came with the Oriental
migrants, or perhaps its ancestors did. But the larger
Northern or Russian Bullfinch (*P. major*) has no doubt
entered our Continent directly from the east. We have
in many groups similar instances of closely allied species
or varieties, one of which, originating at a somewhat
later stage than the other, took a different route of
migration from that followed by its near relative.

The Pine-Grosbeak (*Pinicola enucleator*) is only
known to British ornithologists as an exceedingly
rare visitor. Its real home lies in the northern parts
of Europe, Asia, and North America, and it is one
of the most typical of the Siberian migrants.

But there are a number of other species of birds,
which, though probably not of Siberian origin,

only migrated westward recently, and have either not yet reached the British Islands, or which lead one to suppose, from their British range, that they are eastern forms.

Such, for instance, is the Nightingale (*Daulias luscinia*), which is probably of Oriental origin, but only visits England regularly in spring. There is no authenticated record of its ever having migrated either to Scotland or Ireland.

The Bearded Titmouse (*Panurus biarmicus*) is one of the eastern birds still resident in England, though unfortunately it seems to be on the verge of extinction. It is unknown in Scotland and Ireland. Another resident eastern species is the Nuthatch (*Sitta cæsia*), but neither of these is probably of Siberian origin.

The majority of the European Reptiles are probably of eastern origin. Among our British species, the Common Viper (*Pelias berus*), for example, is a typically eastern form. It is almost unknown in Southern Europe proper—that is to say, in Italy, the Balkan peninsula, and the Mediterranean Islands, but its range extends in the west as far as Spain, and in the east right across the Asiatic continent to Japan. It is well known that the Viper occurs in Scotland, and that neither it nor any other snake is found in Ireland. There is a legend, indeed, that snakes did once exist in Ireland and were banished from the island by St. Patrick, but unfortunately we have no historical evidence that such an interesting event actually took place. The Sand-Lizard (*Lacerta agilis*), another British species, may be looked upon as an eastern form. It is quite absent from Italy, the Balkan peninsula, and the Mediterranean Islands, but extends throughout Central Europe to the east.

Among the species of eastern Reptiles which have a mere local range in Europe might be mentioned the two Lizards, *Phrynocephalus auritus* and *Agama sanguinolenta*. They belong to the family *Iguanidæ*, which includes some very large species. Both of them are Asiatic forms, which have only just penetrated across the eastern steppes into Europe, where they inhabit the arid regions between the Caspian and the River Don in Southern Russia.

The species of Mammals living in Europe at the present day have, with few exceptions, migrated to our continent from other parts of the world. With regard to the Birds, it is possible that a somewhat larger number proportionally may be of European origin. Still, the great majority are, I think, to be regarded as immigrants. The autochthones are about equal to the immigrant reptiles, but many of the European Amphibians and the majority of the Fishes have probably originated on our continent. Some of the European Amphibia—especially among the tailless forms—appear to be immigrants from Asia. Thus the distribution of *Rana arvalis* in Europe is remarkably like that of a Siberian migrant. This frog occurs in Siberia,

13

ranging southward as far as Persia and parts of
Asia Minor. Crossing the European border, we
find it in Russia, Upper Hungary, North and Central
Germany,—being rarer in the south,—Denmark, and
Scandinavia. According to Bedriaga, it crosses the
Rhine only in Alsace, but occurs no farther west. It
only just enters Holland. If we suppose the species
to have originated in Central Europe, we should
expect to find it in Switzerland, France, and perhaps
England. If it had its ancestral home in Eastern
Europe, we might expect it to occur on the Balkan
peninsula. It seems to me more probable, therefore,
that *Rana arvalis* came with the Siberian migration.
This need not cause surprise, as the genus *Rana* is
certainly not European. Out of about 110 species,
only four are peculiar to Europe, the rest are scattered
over all parts of the globe. Moreover, the fact that
these four species are confined to Southern Europe
would seem to indicate that the first species entered
from the south, and there either became modified or
spread over nearly the whole continent, as did, for
instance, *Rana esculenta* and *R. temporaria*. Neither
of these is by any means confined to Europe. *R.
esculenta* ranges right across the Asiatic continent to
Japan, and also enters North Africa, while the other
has a wide distribution in northern and temperate
Asia.

The various groups of Vertebrates are not dependent
on each other in their migrations. Mammals and Birds
extend their range with so much greater facility than

Reptiles and Amphibians, that the surplus population of our neighbouring continents readily poured into Europe when—owing to changes of climate perhaps—they forsook their original homes.

We observe much the same differences of origin in the various groups of European Invertebrates. The Central European Molluscan fauna, remarks Dr. Kobelt, had already developed from the pliocene —in almost all its details, as regards formation of species and distribution—when the Ice-Age commenced (*b*, i. p. 162). Certain very interesting dislocations, however, in the range of land mollusca can be proved to have taken place about that time. Thus, as Dr. Kobelt has pointed out, the genus *Zonites*, which is now almost confined to the south-east of Europe, occurs in inter-glacial deposits in the valley of the Neckar, and even as far west as the Seine. If we might judge from this single instance, a molluscan migration from the east to the west seems to have occurred either in early or pre-glacial times. That *Helix pomatia* has migrated only comparatively recently from the East to Western Europe is rendered probable by its general range in northern and western Europe, but I cannot agree with Dr. Kobelt in the belief that *Helix aspersa* is of an equally recent origin in the North. No matter whether it has been found fossil or no, its range in the British Islands points to its having penetrated to Ireland when the latter was still connected with the Continent by way of England. Its migration from

the Mediterranean dates therefore from early pleistocene or late pliocene times.

In referring to the sixty-five species of Land and
Freshwater Mollusca which have been described from
the continental "Loess," Dr. Kobelt states (p. 166)
that this fauna has certainly not a steppe-character.
It does not therefore strengthen Professor Nehring's
view that Europe during the deposition of the loess
had a climate comparable to that of the Siberian
steppes. The Glacial period had hardly any effect
on the molluscan fauna of Europe. Dr. Kobelt
believes in a certain movement of that fauna from
the least favourable areas, with a subsequent
re-immigration; but even that could not have
taken place on a large scale. Nothing like a
destruction of the fauna occurred, as far as we
know from fossil evidence.

Not a single species of land or freshwater mollusc
can be quoted as having migrated to Europe from
Siberia in recent geological times. The molluscan
fauna of the latter country is so closely connected
with that of Europe, that it is quite impossible to
elevate it to the rank of a sub-region of the Holarctic
Region. Dr. Kobelt insists that Siberia cannot even
claim to be placed into a distinct province. According to the same authority, we find no species in the
whole Siberian molluscan fauna which we might
regard as having immigrated since the close of the
Glacial period. Even to attempt the location of
the original homes of many of the species which

Siberia has in common with Europe, seems hopeless. Such forms as *Arion hortensis*, which has been obtained in Siberia, and which, as we have seen, must have originated in Western Europe, migrated in pliocene or miocene times, possibly along the shores of the Mediterranean and across Asia Minor. We have evidence, therefore, of an eastward migration among the land and freshwater mollusca in later Tertiary times, but not of a westward one from Siberia.

A very different view is presented to us by the coleopterous fauna of Europe. Many of our European Beetles are Siberian migrants. Let us take, for instance, the Tiger Beetles (*Cicindelidæ*). There are over forty species of the genus *Cicindela* in Europe, five of which reach the British Islands. This seems a large number; but there are altogether no less than 6co species of the genus scattered over the greater part of the world, many of them being Asiatic. The genus is certainly not of European origin, for not only are most of the European species confined to the Caucasus and the south-east generally, but no *Cicindelidæ* whatsoever occur, for example, in Madeira or the Canaries, where we should expect some to have persisted if the genus had originated on our continent. Moreover, of the five tribes into which the large family of *Cicindelidæ* can be sub-divided, only two range to Europe, and one of them is represented by only a single species on our continent.

Some of the *Cicindelas* may have come with the
Oriental migration. I think this was the case with
the only Irish species of the genus, *C. campestris.* It
occurs all over continental Europe and Northern Asia,
and varieties of the species are known from Corsica,
Sicily, Crete, the Cyclades, Sardinia, Asia Minor,
Greece, and Spain. Five species of *Cicindela*, as I
said, are known from England, of which *C. silvatica*
and *C. maritima* are certainly Siberian migrants, and
perhaps *C. hybrida* too. Neither of the two first
species is found in Southern Europe or in Spain,
where we should expect them to occur had they
originated on our continent. *C. silvatica* and
maritima have no doubt entered Europe from
Siberia in recent geological times, probably soon
after a way was opened up across the Tchornosjem
district of Southern Russia—that is to say, in
inter-glacial times. The former spread along the
Central European plain as far west as the south-east
of England when Great Britain still formed part of
France. *C. maritima*, which preferred the proximity
of the sea, migrated along the shores of the Caspian
and then across Russia to the shores of the Baltic
and North Sea, and has penetrated a little farther
north and west in England than its near relative.
C. litterata has a very similar distribution and origin,
but instead of wandering so far west as the British
Islands, it seems to have preferred extending its
range southward, and has just reached Northern
Italy.

The closely allied Ground-beetles (*Carabidæ*) furnish us with equally interesting and instructive proofs of a migration from Asia. Over 300 species of *Carabus* are known to science. The number of species inhabiting Asia and Europe are about equal. But the genus does not extend its range to Southern Asia or to South America or Australia. Very few species enter Africa, and only nine North America, of which three also occur in Siberia. The genus is unknown in Madeira, and only represented by three species in the Canary Islands. To judge from its distribution, it has probably originated in Western Asia. Probably some *Carabi* of European origin have spread into Asia, but the Asiatic—or we might say the Siberian—origin and subsequent migration westward of a number of well-known forms appears to me evident. Such forms as *C. clathratus*, *C. granulatus*, and *C. cancellatus* are no doubt of European origin, and have only in recent geological times extended their range across Northern Asia, whilst *C. marginalis*, coming from Siberia, can hardly be said to have invaded Europe, since it has never been met with farther west than the eastern provinces of Prussia.

Among the *Carabidæ* there are altogether very many examples pointing to a migration from Asia to Europe, but I do not wish here to give a list of all such cases, and only refer to a few of the more remarkable ones. One of the European species of Demetrias (*D. unipunctatus*), known to English entomologists as a south-eastern form, seems to have arrived with the Siberian migration, whilst the closely allied *D. atricapillus*, which has been able to reach Ireland, has a wider range and came earlier with the Orientals.

Messrs. Speyer state (p. 68) that almost all those species of Central European Butterflies whose northern limit is deflected southward as we approach the west coast of Europe, inhabit also the Volga country and the adjoining parts of Asia. Many of them are much commoner there than in Central Europe, and it appears probable to the authors of the *Geographical Distribution of Butterflies* that these species came from the east. Asia and Central Europe have, according to Messrs. Speyer, no fewer than 156 species in common. Mr. Petersen estimates that no less than 91 per cent. of the Arctic-European Butterflies also occur in Siberia. He made a special study of the Arctic *Macro-lepidoptera*, and came to the conclusion that Central Asia, not having been glaciated in the Ice-Age, offered a possibility of existence to both animals and plants. Here, he thinks, was the principal centre to which Europe owed its re-population in post-glacial times. Mr. Petersen is of opinion (p. 40) that the Arctic-European *Lepidoptera* are composed of two elements — the pliocene relics which persisted in Europe during the Glacial period, and the new immigrants from Siberia.

No doubt Siberia supplied Europe with a number

of species of Butterflies and Moths in recent geological times, but we need not necessarily suppose
that these arrived only after the Glacial period. Even
the most extreme glacialists admit that large areas
on our continent were free from ice at the height of
the Ice-Age, Siberia had therefore no particular
advantage over Europe in giving an asylum to
Butterflies and Moths which were escaping from the
rigours of a supposed arctic climate. But we have
already learned (p. 80) that the climate during the
Glacial period probably differed but little from that
which we enjoy at the present day, and we may
assume, therefore, that the *Lepidoptera* of Siberia
migrated during that time or even earlier to
Europe.

Let us for a moment reconsider some instances of
mammalian migration from Siberia, with a view to
studying more closely the nature of these great
events. I mentioned the fact that some of the
Siberian migrants have remained in England, that
more have settled down permanently on our continent, but that many others have either become
entirely extinct or do not live any longer in
Europe.

Of the mammals which made their appearance in
Great Britain in recent geological times, *i.e.*, during
and since the deposition of the Forest-Bed for example,
the following species probably came direct from
Siberia across the plains of Europe, as already mentioned (p. 95):—

Canis lagopus.

Gulo luscus.

* Mustela erminea.

* „ putorius.

* „ vulgaris.

* Sorex vulgaris.

Lagomys pusillus.

* Castor fiber.

Spermophilus Eversmanni.

„ erythrogenoides.

Cricetus songarus.

Myodes lemmus.

Cuniculus torquatus.

* Mus minutus.

* Arvicola agrestis.

* „ amphibius.

„ arvalis.

* „ glareolus.

„ gregalis.

„ ratticeps.

Equus caballus.

Saiga tartarica.

Ovibos moschatus.

Alces latifrons.

„ machlis.

Rangifer tarandus.

* Those marked with an asterisk still inhabit Great Britain, or did so within historic times.

Of the arrival of many of these in Europe we have geological proof, as they have left their bones in recent pleistocene deposits, and are unknown from older European strata. The remote ancestors of others, such as *Sorex* and *Lagomys*, no doubt lived in Europe, but the recent species probably had their original homes in Asia. It is evident that in recent geological times there existed no active centre of origin for mammals in Europe, and that our continent was largely dependent on the neighbouring one for the supply of its mammalian fauna. A shifting of the centre of development from Europe to Asia appears to have taken place occasionally, as already mentioned (p. 45). Mr. Lydekker has drawn attention to the fact that though the remote ancestors of the *Elephantidæ* resided in Europe, neither the latter

continent nor North America was the home of the
direct ancestor of any of the true Elephants.
Similarly, though we have had our *Sorex* in Europe
from' the Upper Eocene and *Lagomys* from the
Middle Miocene, the geographical distribution of
Sorex vulgaris and *Lagomys pusillus* does not support
the view that they are of European origin and have
migrated to Asia. Their absence from most of
the European islands indicates either an extremely
recent origin or a recent immigration from Asia,
and the latter view seems to me much the more
probable.

No less than twenty-six species of the Siberian
mammals penetrated as far west as the British Islands,
and nine of these still inhabit Great Britain. Some
of the remaining seventeen species probably lived only
for a very short time in England, and the rest
gradually became extinct one by one. This process
of extinction of the aliens still continues. The Beaver
(*Castor fiber*) has died out within recent historic times.
We possess legends and uncertain historic records
pointing to the existence of the Reindeer in Scotland
as recently as about seven centuries ago. But much
the same state of things has happened on the Continent. The Glutton (*Gulo luscus*), which still lived
in Northern Germany last century, has now entirely
vanished from that country, as also the Reindeer.
The Lemmings have found an asylum in Scandinavia.
The Musk-Ox (*Ovibos moschatus*) has disappeared
not only from Europe but also from Asia, and is now

confined to Arctic America and Greenland. The Horse no longer occurs in Europe in the wild state, and the Saiga Antelope (*Saiga tartarica*) has retreated to the Steppes of Eastern Europe and Western Siberia.

As we proceed more and more eastward across Central Europe, we find that a larger and larger percentage of the Siberian migrants have adopted the new country as their permanent home, though in France and Germany, as well as in Austria, we have evidence that a great number of Siberian species, which formerly lived there, have either become entirely extinct, or have retreated towards the land of their origin. There is a prevalent belief that these migrants have taken refuge on the higher European mountain ranges, but this idea is altogether erroneous, as will be shown in the chapter dealing with the origin of the Alpine fauna.

One of the Jerboas (*Alactaga jaculus*) occurs fossil as far west as Western Germany, but it is now confined to Russia and Western Siberia. The Bobak marmot (*Arctomys bobak*), which has a similar range now, probably inhabited France in former times. A Siberian species which has retreated but little is the Hamster (*Cricetus vulgaris*). Its fossil remains have been found in Central France, but it does not now occur west of the Vosges Mountains.

It appears, therefore, as if a wave of migration had swept over Central Europe from east to west, that those species which were able to adapt themselves to the new surroundings had remained, and as if the rest had died out or were gradually retreating to the east,

Ornithologists are well acquainted with the fact that in some years there is an unusually large exodus from Eastern Europe and Siberia of birds; and that species like the Waxwing (*Ampelis garrulus*) then appear in great numbers. But the appearance of this bird in Western Europe is not looked upon as so remarkable as that of Pallas's Sandgrouse (*Syrrhaptes paradoxus*, Fig. 3, p. 42), a typical inhabitant and resident of the Arctic Steppes. The last great irruption took place in 1888, and many birds reached even the extreme west of Ireland in May and June of that year. A few weeks before, it had been announced to the German papers that large flocks of this peculiar pigeon-like bird had arrived in the eastern provinces; and though the vast majority vanished as quickly as they had come, a certain number remained for a year or so in the newly visited countries, and some even bred in England.

Twenty-five years before, in 1863, a similar migration had occurred, though not perhaps on quite such a vast scale, and a few small flocks had made their appearance in Western Europe on several occasions between these dates.

It may not be generally known that no other bird has been honoured by our Government in a like manner, for it is the only animal for whose protection a separate Act of Parliament has been passed.

In spite of this unusual precaution, the species has
not survived to add another member to the resident
British fauna. The wave of migration from the east
has come and vanished again just like so many others
with which history is familiar.

These migrations from the east occurring at the
present day give us some idea of those of which we
have fossil evidence, and which all had their origin
in Central and Northern Asia. Almost all the species
of mammals to which I have referred as being of
Siberian origin have been found in the fossil state
in comparatively recent geological deposits within a
certain very limited area. None of the typical species
have ever been found in Southern Europe proper,
including the Mediterranean islands. It must be
remembered that though the Reindeer is a Siberian
migrant, the form of the Reindeer which was found
in the Pyrenees belonged to a distinct variety—in
fact, to a much earlier migration which issued from
the Arctic European Regions, and to which I have
referred in detail (pp. 150-158). Curiously enough,
no deposits of these typical Siberian mammals
have ever been obtained in Scandinavia—only in
Russia, Austria, Switzerland (the lowlands), Germany,
Belgium, France, and England. To facilitate a study
of the extent of these migrations, I have constructed
a map on which the probable course taken across
Central Europe is roughly indicated by dots
(Fig. 16).

In the migrations of to-day we perceive the same

tendency as in the older ones of which we have fossil
evidence, viz., generally a spreading of species on a
large scale over new territory, and then a gradual

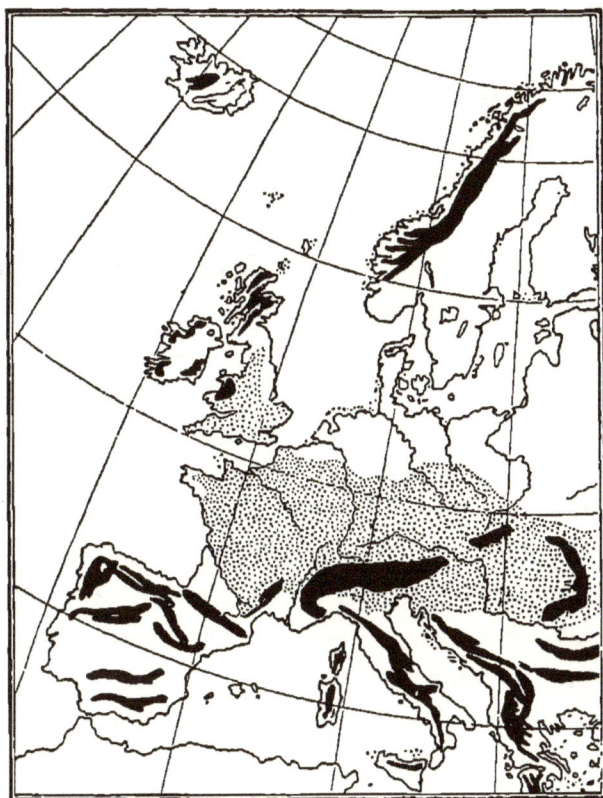

FIG. 16.—Map of Europe.   The dotted portions represent, approximately, the course of migration of the Siberian mammals.   The
principal mountain ranges are roughly indicated in black.

shrinkage towards their original home, with an occasional survival of small colonies in the invaded part.
It must not be supposed that this observation applies

alone to the Siberian migration. In the case of the
Arctic one, precisely the same thing has happened,
and we shall see that the Southern (migration from
the south) agrees in this respect with the others.

As for the immediate cause of these migrations, it is
to be looked for either in the scarcity of food dependent
upon a temporary or permanent change of climate, or
in an excessive increase in numbers of a particular
species. I do not propose to trace back migrations
beyond the Pliocene Epoch, or indeed much beyond
the beginning of the Glacial period, which is regarded
as a phase of the most recent geological epoch, viz.,
the Pleistocene. During the period in question, we
have indirect evidence of one vast migration from
Siberia into Europe across the lowlands lying to the
north of the Caspian and to the south of the Ural
Mountains. There is a general consensus of opinion
that this migration took place in Pleistocene times.
Professor Nehring thinks that there can be no doubt
(p. 222) that the Siberian migrants arrived in
Northern Germany after the first stage or division
of the Glacial period, and lived there probably
during the inter-glacial phase which occurred between the first and second stages—if indeed we look
upon this period as being divisible into two distinct
stages.

Judging from the evidence of distribution of
mammals in pleistocene Europe, Professor Boyd
Dawkins came to the conclusion (p. 113) that the
climate of our continent " was severe in the north

and warm in the south, while in the middle zone, comprising France, Germany, and the greater part of Britain, the winters were cold and the summers warm, as in Middle Asia and North America." "In the summer time the southern species would pass northwards, and in the winter time the northern would swing southwards, and thus occupy at different times of the year the same tract of ground, as is now the case with the Elks and Reindeer." Very different are the views of Professor Nehring on this subject. According to him, the climate in Germany must have been extremely cold and damp, resembling that of Greenland, though perhaps not quite so arctic. Professor Nehring does not at all believe that southern and northern species of mammals could have lived in Central or Northern Europe at the same time; though of this we have undoubted geological evidence (pp. 72-75). He thinks that the supposed commingling of southern and northern types, which has actually been shown by Professor Dawkins to occur, is either due to careless observation or to the fact that some of the species need not necessarily have lived where their bones were found (p. 133).

The most reliable conclusions as regards former conditions of vegetation and climate can be drawn, according to Professor Nehring, from the smaller burrowing mammals, such as the marmots, sousliks, etc. He is of opinion that a great portion of Northern Europe, where their remains have been

14

discovered, must have possessed tundras and steppes, as we find them nowadays in Siberia, and a climate similar to that of Northern Asia. It is presumed that the climate, after the maximum cold of the first stage of the Ice-Age, ameliorated so far as to permit these mammals to exist in Europe.

The natural question, however, which is forced upon us in reading Professor Nehring's interesting and suggestive work is, where did all these steppe animals live during the earlier part of the Ice-Age? No traces of their remains have been discovered in Southern Europe, and it can therefore certainly be affirmed that they could not have lived there. If Central and Northern Europe were uninhabitable for mammals, Central and Northern Asia must have been even more so, and we have to fall back upon the Oriental Region as a possible home of these species during the assumed maximum cold of the Glacial period. In invading Europe from the Oriental Region these Siberian mammals would have taken the shorter route by Asia Minor and Greece, which was open to them. This they certainly did not do, which proves that they came directly from Siberia to Europe without retreating first to Southern Asia.

But it seems to me that there is no necessity for assuming such drastic changes of climate to have taken place at all (compare pp. 75-80). We really have no idea under what precise climatic conditions the Siberian mammals lived in their original home. The only thing we can be certain of is that the

smaller burrowing mammals would not have chosen
a wood to live in, if they could possibly help it.
Prairies, or sand-dunes with short grass or shrubs,
such as abound in Europe near the sea-coast, would
suit these species perfectly. If we suppose Northern
Germany to have been covered by sea (p. 156) during
part of the Pleistocene Epoch, forests would probably
not have grown there for a very considerable time
afterwards, owing to the excessive salinity of the soil,
but a tract of sandy country would have been left
on the retreat of the sea. Possibly a slight change
of climate in the original home of these steppe-species
may have reduced their habitable area, and thus
caused their migration into Europe.

But this migration problem cannot be solved
without tracing the mammals to their place of
origin and investigating their early history. This I
shall attempt to do presently; meanwhile, it would
be interesting to note whether other groups of
animals support Professor Nehring's steppe-theory.

Among groups other than mammals, the most
important, for the purpose of drawing conclusions
as to former physical conditions and climate, are the
mollusca. Their remains have been well preserved,
and are easily identified. Though Professor Nehring
argues that the molluscs found along with the small
mammals harmonise perfectly with the assumption
of a steppe-climate (p. 212), I cannot at all agree
with him. He enumerates the following sixteen
species as having been discovered by him:—

1. Pupa muscorum.	9. Helix pulchella.
2. Chondrula tridens.	10. Do. hortensis.
3. Cionella lubrica.	11. Do. obvoluta.
4. Patula ruderata.	12. Hyalinia radiatula,
5. Do. rotundata.	13. Succinea oblonga.
6. Helix striata.	14. Limnæa peregra.
7. Do. hispidia.	15. Clausilia sp.
8. Do. tenuilabris.	16. Pisidium pusillum.

Only two of these can be looked upon as typically
northern species, viz., *Patula ruderata* and *Helix
tenuilabris*, though both of them are still found
living locally in Germany. Some of the others
are decidedly southern species, like *Chondrula tridens*,
Helix obvoluta, H. rotundata, and *H. striata*. All the
rest live and flourish, for example, in Ireland at the
present day, where, as we all know, anything but a
dry steppe-climate prevails.

Dr. Kobelt quite agrees with me in thinking that
the remains of the mollusca found along with the
so-called " steppe-mammals " afford no proof of a
steppe-character of the country at the time when
they were alive (p. 166). Nor do the mollusca which
have been found in England in the Forest-Bed and
the succeeding pleistocene strata support such a
view. The Forest-Bed, generally regarded as belonging to the Upper Pliocene, I believe to be an inter-
glacial pleistocene deposit—contemporaneous with the
loess formation in Germany. Of fifty-nine species
of land and freshwater mollusca which have been
discovered in this bed, forty-eight species, according

to Mr. Clement Reid (p. 186), are at present living in Norfolk, six are extinct, two are continental forms living in the same latitudes as Norfolk, and the other three are all southern forms. Not a single species has a particularly northern range. Of the land and freshwater mollusca of the South of England in the succeeding pleistocene deposits, six species are now no longer living in the British Islands, but only one (*Helix ruderata*) can be looked upon as an Arctic or Alpine form. After this short digression on the mollusca, I will briefly recapitulate what is known about the early history of the Siberian mammals, which will assist us in tracing the cause of their migration to Europe.

We have in Siberia problems quite as difficult of solution as the European ones. Volumes have been written to explain the former presence of Arctic mammals like the Reindeer in Southern Europe, and the most extraordinary demands on the credulity of the public have been made by some geologists in their attempts to account for this comparatively simple problem. In Northern Asia a somewhat similar phenomenon, but much more difficult of explanation, has taken place. Mammals have been found fossil in recent geological deposits in localities where they do not now occur, and apparently the Siberian and the European deposits are of about the same age. Now, however, comes the extraordinary difference. In Europe the Arctic mammals went southward, but in Siberia the Southern ones went northward. Not

only do we find the Saiga-Antelope, Tiger, Wild
Horse, European Bison, Mammoth, and Rhinoceros
in the extreme north of the mainland of Siberia; their
remains have even been obtained in the New Siberian
Islands. As these islands are situated in the same
latitude as the northern part of Novaya Zemlya,—
indeed, not far south of the latitude of Spitsbergen,
—the fact of such huge mammals having been able
to find subsistence there at apparently quite a recent
geological period seems an astounding fact. It may
be urged that their bones might have been carried so
far north by ice, or by some other equally powerful
agency. But Tcherski and all other palæontologists
who have examined these northern deposits are
unanimous in the belief that these herbivores and
carnivores lived and died where their remains are
now found. " It is evident," says Tcherski (p. 451),
"that these large animals could only have lived in
those extremely northern latitudes under correspond-
ingly favourable conditions of the vegetation, viz.,
during the existence of forests, meadows, and
steppes." He also is of opinion that the moist climate
which evidently prevailed in Europe during Post-
tertiary (Pleistocene) times must have modified the
Siberian climate in so far as to render it milder.
The existence of the Aralo-Caspian basin (Fig. 12,
p. 156) must also have tended in the same direction.
It appears then that, at the time when plants and
animals are believed to have retired southward in
Europe before the supposed advancing Scandinavian

ice-sheet, no agency existed in North Siberia which
was able to suppress and to annihilate the forest and
meadow vegetation, and drive away the fauna connected with it. We know, continues Tcherski, that
such genera as Bison, Colus (*Saiga*), Rhinoceros,
Elephas, and Equus are met with in all horizons
of the diluvium of West Siberia. He therefore
comes to the conclusion (p. 474), that these and
other facts imply that the retreat of the North
Asiatic fauna commenced about the end of the
Tertiary Era (Pliocene), and that it was continued
very slowly throughout the Post-tertiary (Pleistocene)
Epoch, without any visible changes in its southward
direction, even *during the time of the most important
glacial developments in Northern Europe.* Only after
the conditions disappeared which had produced the
augmentation of an atmospheric moisture, did the
climate of North Siberia become deadly to a
temperate fauna and flora. Tundras then spread
over the meadow-lands and remnants of forests,
whilst arctic animals replaced the large ungulates
and carnivores which had wandered far away from
their native southern home.

This is Tcherski's explanation of the extraordinary
events which he has chronicled, after years of the
most arduous labour and under conditions of peculiar
hardship. And though his work cannot be over-
estimated, and his opinions should receive the most
careful consideration, yet I fear the explanation will
not be looked upon as entirely satisfactory. Every one

will agree with him that the climate of Siberia must
have been greatly moister in pliocene and pleistocene
times than it is now. The Aralo-Caspian covered
a vast area of South-western Siberia. Freshwater
basins existed along the east of the Ural Mountains,
while Central Asia was studded over with a number of
large lakes, which have now almost entirely vanished.
But that the generally assumed refrigeration of
Europe must have had a chilling effect on the
Siberian atmosphere seems to me evident. That the
whole of Northern Europe should have been made
uninhabitable owing to the advance of the Scandinavian ice-sheet, while North Siberia at the same
time supported forests, meadows, and a temperate
fauna, is incredible. At the approach of winter, at
any rate, the animals would have been driven southward for thousands of miles to seek shelter from the
snows and cold and to obtain nourishment, and it
would scarcely have been possible for them to
undertake such vast migrations at every season.
Professor James Geikie's suggestion (p. 706), that
the Mammoth and Woolly Rhinoceros could have
survived the Pleistocene Epoch in Southern Siberia,
does not appear to solve the problem, as that part
of Asia must have participated in the great cold
which is said to have prevailed all over Europe.

Let us now concede, for the sake of argument,
that the current views regarding the pleistocene
climate of Europe are correct. We are told by Professor Geikie that the climate of Scotland during

part of the Pleistocene Epoch was so cold, that the
whole country was buried underneath one immense
mèr de glace, through which peered only the higher
mountain-tops (p. 67). If this was the state of climate
in close proximity to the Atlantic, it must probably
have been still more severe on the European continent. Now at the present time Siberia has the
reputation of being the coldest country in the world,
and the mercury of the thermometer is said to remain
frozen for weeks during winter, even in the south.

With the prevailing dampness in pleistocene times
the snowfall throughout Siberia would have been
much heavier than at present, though it would have
modified the temperature to some extent. Under
such circumstances Southern Siberia could not have
been a desirable place of residence for large mammals.
It would have been necessary for the Mammoth and
the other species referred to, to wander farther into
the extreme south of Asia or Europe to find a
suitable refuge during the arctic conditions which
are supposed to have prevailed in Northern Europe.
To quote Professor J. Geikie's own words (p. 706):
"They (Mammoth, etc.) would seem to have lived in
Southern Siberia throughout the whole Pleistocene
period, from which region doubtless they originally invaded our Continent. But with the approach of our
genial forest-epoch (penultimate inter-glacial stage)
they gradually vanished from Europe, to linger for a
long time in Siberia before they finally died out."
It is suggested, therefore, by the author that the

Mammoth and the other mammals whose remains have been discovered on the New Siberian Islands found their way there during one of the late inter-glacial stages of the Ice-Age. But there is no astonishment expressed by Professor Geikie at the extraordinary change of climate which must have occurred in Siberia to allow of such migrations. I can find no very definite statement in this author's work as to the nature of the climate in Europe during those inter-glacial phases, but he remarks (p. 129) " that the evidence of the Scottish inter-glacial beds, so far as it went, did not entitle us to infer that during their accumulation local glaciers may not have existed in the Highland valleys." There is no evidence, in other words, of the existence in Europe of a milder climate than that prevailing at present. Still less can there be any ground for the supposition that the climate of the whole of Siberia ameliorated to such an extent that forests and meadows could develop as far north as the New Siberian Islands ; for if the temperature in Europe was then about the same as now, that of Siberia could not have been vastly higher than it is at present.

It is highly improbable, therefore, that a sufficiently mild climate prevailed in the extreme north of Siberia during the so-called *later inter-glacial periods* to induce the mammals to which I have referred to seek fresh pastures there.

The late Professor Brandt, one of the highest zoological authorities in Russia, was of opinion that at the commencement of the Glacial period the great mammals of Northern Siberia either perished or migrated southward. From there they gradually penetrated into European Russia. He believed that before the Glacial period a connection existed between the Aralo-Caspian Sea and the Arctic Ocean, carrying warm water northward. The gradual disappearance of this marine channel caused a decrease of warmth in Northern Asia, so that large accumulations of frozen soil and ice were formed, which still more depressed the temperature. This, he suggested, probably took place at the time when the Glacial period commenced in North-western Europe.

It has been urged against these views of Tcherski and Brandt, that the bone beds in the Liakov Islands (New Siberian Islands) rest partly upon a solid layer of ice of nearly seventy feet thick. This mass of ice, it was thought, must have accumulated during the Glacial period. As the bones rest upon it, the mammals could only have lived in those islands in more recent times, after the Ice-Age had passed away. Nothing, apparently, can be clearer, and yet in the face of this seeming proof one feels, as I have mentioned before, that if such an extraordinary revolution of climate as is implied by this admission had taken place, we should be able to perceive the traces throughout the northern hemisphere. In this dilemma, a suggestion made by Dr. Bunge, who visited the New Siberian Islands recently at the instance of the Imperial Academy of St. Petersburg, helps us out of the difficulty. He found that, as a rule, these so-called fossil glaciers contain seams of mud and sand, and he argued that the ice had formed, and is still forming at the present day, in fissures of the earth. I entirely concur with this view. Neither palæontology nor the geographical distribution of animals lend any support to the other theory, and I think we may conclude that Brandt's view in the main is probably the correct explanation of the phenomena which we have discussed. Some important facts of distribution are more easily explicable on this assumption. Why, for instance, should the Siberian fauna of pliocene times have remained in Siberia and not have migrated to Europe at that time? The pliocene mammals of Siberia are mostly of southern origin. Their range increased enormously during the epoch throughout Northern Asia. We should expect them, therefore, to have crossed the Caspian plains, or even the low-lying Ural Mountains, to pour into the neighbouring continent. But Professor Brandt explained how they were prevented from spreading west. An arm of the sea stretched from the Aralo-Caspian to the Arctic Ocean, thus raising an effectual barrier between the two continents. There is some evidence for the belief, as we shall learn presently, that this marine barrier existed also during the early part of the pleistocene epoch. After having greatly expanded during pliocene times, the fauna of Siberia gradually withdrew from the

northern regions during the earlier portion of the
succeeding epoch. It was only after the marine
connection above referred to ceased to exist, or
became disconnected, that an entry into Europe
was possible.

A fauna, to some extent composed of species now
inhabiting the steppes of Eastern Europe and Siberia,
poured into the neighbouring continent. On p. 95 I
have given a list of those which reached as far west
as the British Islands, but, as I mentioned, many
other species came from the east about this time.
With regard to the early history of the Siberian
mammals, I favour a view somewhat between that
of Tcherski and that of Brandt. The outpouring of
the fauna into Europe seems to me to indicate that
there was a sudden change of climate in Siberia.
This was produced, perhaps, by the rupture of the
marine connection between the Arctic Ocean and
the Aralo-Caspian. Such an event would not only
have caused the sudden shrinkage of the area available for food-supply by lowering the temperature in
Siberia, it would have acted also as a means in
assisting the fauna to enter a new continent where
an inconsiderable number of mammals, already established, were mostly dispossessed of their homes by the
advancing eastern host.

Brandt's theory, however, of a marine connection
between the Arctic Ocean and the Aralo-Caspian is
by no means generally accepted. That the Caspian
Sea was at that time greatly larger than it is at

present, and joined to the Sea of Aral and the Black Sea, is acknowledged by everybody. That the deposits laid down by this huge inland sea reach as far north as the shores of the river Kama, in Central Russia, is also well known to geologists. But what comes rather as a surprise, is that Professor Karpinski, whom we must take as one of the highest authorities on the geology of Russia, asserts that this Aralo-Caspian Sea was probably joined by a system of lakes or channels to the Arctic Ocean (p. 183). He was by no means the first, though, to put forward such a theory. We have already learned that Professor Brandt held a somewhat similar view, though he believed in something more than a connection by mere channels, and Mr. Köppen, and also the Russian traveller Mr. Kessler, agreed with him. So much was Professor Boyd Dawkins impressed with their arguments at the time, that he wrote (*c*, p. 148): "Before the lowering of the temperature in Central Europe, the sea had already rolled through the low country of Russia, from the Caspian to the White Sea and the Baltic, and formed a barrier to western migration to the Arctic mammals of Asia."

In one particular Professor Dawkins's views differ from those of almost all the previous writers. His connection between the Caspian and the Arctic Ocean is placed to the west of the Ural Mountains, while it had always been assumed by the Russian writers to have lain on the eastern or Asiatic side of that mountain range. Thus, when Tcherski in recent

years announced that the tract on this eastern side of
the mountains was covered by freshwater deposits,
his discovery seemed once for all to settle the problem
of the arctic marine connection in the negative. As
Professor Dawkins's theory has, however, received
much additional affirmative evidence by current
faunal researches, a connection between the Caspian
(or Aralo-Caspian) and the Arctic Ocean (White Sea)
may have actually existed within recent geological
times.

What *relict lakes* are, has already been explained
(p. 176), and their fauna will again be referred to in
a subsequent chapter. I might perhaps be allowed
to repeat that such lakes are supposed to have been
flooded by, or to have been in close connection with,
the sea at some former period. Many of the Swedish
lakes are spoken of as relict lakes (Reliktenseen),
because they contain a number of marine species of
animals which have now become adapted to live in
fresh water, but all of whose nearest relatives inhabit
the sea. One of these, the schizopod crustacean *Mysis
relicta*,—a shrimp-like creature,—which was formerly
believed to inhabit also the Caspian, is of particular
interest. More recently, the occurrence of this *Mysis*
in the Caspian was denied, but though this denial has
been confirmed by Professor Sars in his memoir on
the crustaceans of the great Russian inland sea, he
has been enabled to add two new species of *Mysis*
to the list of those already known to science. These
are *M. caspia* and *M. micropthalma*, and both are

closely related to the arctic marine *Mysis oculata*. According to Professor Sars, the genus *Mysis* as a whole may be regarded as arctic in character. The occurrence of these two species, therefore, in his opinion, points to a recent connection of the Caspian with the Glacial Sea.

A large number of other crustaceans have been described by the same author from the Caspian. Of the order Cumacea, which is exclusively marine, ten species are mentioned, but none of these seems to range beyond the Caspian. Among the smaller species of crustaceans, a minute pelagic copepod (*Limnocalanus grimaldii*) also inhabits the Baltic and the Arctic Ocean. The marine isopod *Idotea entomon*, related to the common wood-louse, has a similar distribution.

Genuine Arctic species of Fishes do not seem to occur in the Caspian, though some, viz., *Clupea caspia, Atherina pontica, Clupionella Grimmi*, and *Syngnathus bucculentus*, are almost certainly the descendants of marine forms.

The Seal of the Caspian (*Phoca caspica*) is closely allied to the Arctic Seal, and its presence alone in that sea indicates that at no very distant date—at any rate since pliocene times—a closer connection with the Arctic Ocean existed than at present.

I am sure it will be readily granted that there is zoological evidence for the belief of such a connection or union between the two great seas. However, it may be urged that owing to the presence of an ice-

sheet in Northern Europe during the Glacial period,
such a connection must either have been pre-glacial
or have existed after that period. But the connection must have occurred at a time when the Caspian

FIG. 17.—Map of European Russia (after Karpinski). The faintly
dotted parts indicate the areas covered by boulder-clay, the
strongly dotted ones those exhibiting Aralo-Caspian and other
post-pliocene deposits.

extended far to the north—when indeed the so-called
post-tertiary Caspian deposits were laid down (Fig.
17). Since the boulder-clay which covers the plain
of Northern Russia is assumed to be the ground-

15

moraine of the great northern ice-sheet, we might
expect to find that the Caspian deposits were not
contemporaneous with it. Curiously enough, it has
been shown by Mr. Sjögren that all observations
have pointed to the fact that these two deposits do
not overlie one another, but occur side by side,
and are therefore contemporaneous. This seems to
warrant our belief, that while the boulder-clay was
being laid down in Northern Europe, the Aralo-
Caspian Sea had some communication with the
White Sea.

The boulder-clay of Northern Continental Europe,
as already stated, is now generally recognised to
be the product of a huge ice-sheet which invaded
the lowlands of Continental Europe from the Scandinavian mountains. Though Alpine glaciers at the
present day produce little or no ground moraine,
these ancient larger ice-sheets, or "mers-de-glace," are
believed to have deposited immense layers of mud
containing scratched and polished stones. Many of
the latter have been carried great distances from
their source of origin. The Scandinavian ice-sheet
is supposed to have advanced as far south as the
line indicated on the map, after which it gradually
retreated. On this point, however, as in almost every
detail connected with the Glacial period, geologists
are at variance. Professor James Geikie maintains,
that there were no less than four Glacial periods,
separated from one another by milder inter-glacial
phases. On the Continent the view of two Glacial

and one inter-glacial period is, I think, more generally adopted.   Professor Geikie's four periods seem
to·me to have originated in a desire to correlate
the British pleistocene deposits with the continental
ones, and at the same time to retain the old view of
the inter-glacial position of the Forest-Bed.   The
two theories agree in so far as that in both the
glacial conditions culminate in a maximum glaciation, followed by a more temperate phase of climate,
with consequent retreat of the ice-sheets, and finally
by a renewed advance of the glaciers.

We are told that there is not the slightest doubt
about it that a marked but gradual decrease of
temperature took place all over Europe either
during the beginning of the Pleistocene or towards
the end of the Pliocene Epoch.

We might reasonably suppose, then, that a similar
climatic effect was produced in Siberia, in consequence of which the fauna would have been
obliged to retreat from the extreme northern latitudes southward.   No doubt great efforts would
have been made by the members of the Siberian
fauna—at any rate by those possessing strong
power of locomotion—to extend their range in
other directions.   But we have no evidence that
a migration from Siberia came to Eastern Europe
at that time.   It seems, therefore, as if the barrier
referred to by Brandt, Köppen, Boyd Dawkins,
and others (p. 222), had existed at this time.   This
would have effectually prevented an overflow of

the fauna from Siberia. Only in deposits later
than the lower continental boulder-clay do we find
traces of a Siberian migration. The time of maximum glaciation had then passed away; the great
glacier which was believed to have invaded the
lowlands of Northern Europe had again retreated,
before the Siberian mammals made their appearance
in Germany.

It has been stated above (p. 226) that while the
Russian boulder-clay was being laid down, the Aralo-
Caspian probably had some communication with the
White Sea.

But how can this view be reconciled with the
existence of a huge *mer de glace* in the northern
plains of Russia? The existence of the ice-sheet
has been conjured up in order to explain the
presence of the boulder-clay. But not long ago a
very different interpretation of the origin of this clay
was given; and one, I may say, which explains the
history of the Siberian and the European fauna in a
more satisfactory manner than is done by the ice-
sheet hypothesis. It is that the boulder-clay is not
the product of land-ice, but has been deposited by a
sea with floating icebergs. Thus the latter hypothesis
does not deny the existence of glaciers, but allows
the mud to be deposited on the floor of a turbid
sea, instead of beneath an immense *mer de glace*. I
need hardly mention that this view, which was
formerly universally accepted by geologists, is now
scouted by almost every authority, both British and

Continental. I should scarcely venture the attempt
to revive old memories and stir up again long
forgotten controversies, were it not for the fact that
many new points have arisen in the course of the
above inquiries, which appear to me so very difficult
to explain by the land-ice hypothesis, while they are
comparatively easy to understand when we adopt
the old theory of the marine origin of the boulder-
clay. But a few geologists even at the present day,
while believing in the land-ice theory, recognise that
the marine hypothesis should have some consideration shown to it. I need only remind glacialists of the
work recently published by Professor Bonney. "The
singular mixture," he remarks (p. 280), "and apparent
crossing of the paths of boulders, as already stated,
are less difficult to explain on the hypothesis of
distribution by floating ice than on that of transport
by land-ice, because, in the former case, though the
drift of winds and currents would be generally in one
direction, both might be varied at particular seasons.
So far as concerns the distribution and thickness of
the glacial deposits, there is not much to choose
between either hypothesis; but on that of land-ice
it is extremely difficult to explain the intercalation of
perfectly stratified sands and gravels and of boulder-
clay, as well as the not infrequent signs of bedding
in the latter."

Now with regard to the land-ice theory, several
serious difficulties present themselves in connection
with the origin of the European fauna. In the first

place, as the climate renders Northern Siberia
almost uninhabitable for mammals at the present
day, how much more severe must it have been
during the time of the maximum glaciation in
Europe. As the then existing fauna was not
driven into Europe, where could it possibly have
survived? Secondly, how can we reconcile the
contemporaneous existence of a great inland sea
(the Aralo-Caspian) containing survivals of mild
Sarmatic times with an immense glacier almost
touching it on its northern shores? How did
one of the most characteristic species of that sea,
Dreyssensia polymorpha, come to make its appearance
in the lower boulder-clay of Prussia and then disappear in the upper? And finally, how are we to
explain the sudden appearance of a Siberian fauna
after the deposition of the lower boulder-clay, except
by the removal of a barrier which had prevented
their egress from Siberia?

If we assume that the continental boulder-clay of
Russia has been formed in the manner so ably
explained by Murchison, de Verneuil, and von Key-
serling, viz., by a sea with floating icebergs, the
temperature of Siberia might have been higher than
at present, and have supported a fauna in more
northern latitudes.

The contemporaneousness of the deposits of this
sea with those of the Aralo-Caspian is also rendered
more intelligible. If we suppose, moreover, the connection between the Aralo-Caspian and the White

Sea (Fig. 12, p. 156) to have existed at this time, we possess an explanation of the method of migration of the Arctic marine species into the Southern and of the Caspian species (*Dreyssensia*) into the Northern Sea.

An inter-glacial phase is believed to have supervened after the deposition of the lower boulder-clay, and it is during this period that the Siberian species first appeared in Central Europe. If we assume then that the retreat of the Northern Sea (Fig. 13, p. 170) opened up a passage for the Siberian fauna, we have in this very fact also an explanation of the extraordinarily large exodus of Asiatic mammals, because the great reduction of the marine area in Northern Europe would have had an important influence in lowering the temperature in Asia. Only a sudden change of climate in Siberia could have brought about the migration of the vast hordes of Asiatic mammals whose remains we find in Central and Western Europe in deposits of that period.

Throughout this work we are made acquainted with facts which bear out the view that the climate during the greater part of the Glacial period was mild rather than intensely arctic in Europe. That a huge ice-sheet could have covered Northern Europe under such conditions appears to me very doubtful. No one can deny, however, that glaciers must have existed during the Glacial period in all the mountainous regions of Central and Northern Europe, though their existence is not incompatible

with a mild climate. Tree-ferns and other tropical
vegetation grow at the foot of glaciers in New
Zealand. We need not even go so far afield, for in
Switzerland grapes ripen and an abundant fauna and
flora thrive in close proximity to some of the well-
known glaciers.

One matter of importance still remains to be considered before concluding this chapter, viz., the
fauna contained in the English geological deposit
known as the "Forest-Bed." This interesting
deposit is exposed at the base of a range of cliffs
on the coast of Norfolk. It is composed of beds
of estuarine and marine origin. The tree-stumps
formerly believed to be the remains of trees *in situ*
have, after more careful examination, proved to be
in all cases drifted specimens. A portion of the
"Forest-Bed" no doubt was laid down in close proximity to a large river, and subject to being periodically flooded by it. It is not absolutely certain, therefore, that all the mammals whose remains occur in
this deposit lived in England or whether only on the
banks of the river farther south. Nevertheless, we
may take for granted that some of them did.
England was at the time connected with France
and Belgium, and for our purpose it matters little
whether they had crossed the Channel or inhabited
those parts of the Continent through which the great
river flowed which sent its alluvial detritus as far as
the plains of Norfolk. All we have to remember
is that certain mammals, which appear to have

originated in Siberia, and of which we have some
evidence that they crossed Central Europe in their
westward course, had now reached the great river
just alluded to, which some geologists believe to have
been the Rhine.

I have had occasion to refer to a number of British
mammals (p. 202)—some of which are now extinct—
which I believe to have migrated across the plains of
continental Europe direct from Siberia. There were
twenty-six species of these Siberian mammals; and no
less than ten of these occur in the Forest-Bed. None
appear in any older British deposit. It is perfectly
clear, therefore, that the Forest-Bed must have been
laid down after their immigration into Europe. They
probably wandered to Western Europe very soon
after crossing the eastern boundaries of our continent; the deposits in which they are found are therefore contemporaneous. But we have learned above
(p. 208), that the beds in Eastern Europe in which the
Siberian mammal-remains are found are more recent
than the lower boulder-clay. As already stated, the
Forest-Bed must also be more recent than the lower
continental boulder-clay, and should be included in
the pleistocene series.

That the Forest-Bed is an inter-glacial deposit has
been urged long ago by various writers. Professor
Geikie regards it as stratigraphically contemporaneous with the peat and freshwater beds below the
lower diluvium of Western and Middle Germany, and
as having been laid down during the first Inter-glacial

Epoch of the great Ice-Age. The fact that no boulder-clay underlies the Forest-Bed seems rather a strong argument against the view of its being an inter-glacial deposit. It lies directly on what is known as the Newer Pliocene Crags. If the Forest-Bed is included in the pleistocene series, as I suggested it should, the crags, or a portion of them, would therefore be equivalent as regards time of deposition to the lower continental boulder-clay. And again, if the lower continental boulder-clay is contemporaneous with the Newer Crags, the latter should also be classed with the pleistocene strata. I can scarcely hope that geologists will be ready to admit such a sweeping change of nomenclature without a protest. I venture, therefore, to explain more fully my reasons for adhering to these unorthodox views.

Let us look once more at the map which I constructed (Fig. 12, p. 156) to elucidate the migration of the Arctic terrestrial species to the British Islands. It will be noticed that one continuous ocean extends from the east coast of England across Holland, Northern Germany, and Russia to the White Sea. At the same time Greenland and Northern Scandinavia, Scotland and Southern Scandinavia, are united by a narrow strip of land, and so are England and France. The waters of the Atlantic and this North European Sea do not therefore intermingle at any point, the two seas being absolutely independent of one another.

Such I assume to have been the geographical condition of Northern Europe during the deposition of
the Red Crag. Arctic mollusca were then brought
to the east coast of England, and boulders were
scattered through the beds laid down on that coast
by icebergs which had been cast off by Scandinavian
glaciers on reaching the sea. Bedded clays which
have yielded arctic shells lie beneath the lower continental boulder-clay on the Baltic coast-lands and on
the coast of the White Sea. According to Professor
Geikie, marine clays on the same geological horizon
reach an elevation of some 230 feet. "It would seem,
then," he says, "that before the deposition of the
lower boulder-clay of those regions the Baltic Sea
had open communication with the German Ocean"
(p. 442). All these clays are evidently deposits of the
same sea. But apart from the fact that the Red Crag
and these Baltic deposits are the oldest of the upper
Tertiary beds containing arctic shells, there is no
evidence that they are contemporaneous.

Overlying the same Baltic deposits comes the lower
boulder-clay, reaching a thickness of several hundred
feet in some parts of Germany. It presents, like the
upper clay, frequent interstratification with well-
bedded deposits of sand and gravel. The scarcity of
marine mollusca, the occurrence of striated surfaces,
and the occasional presence of so-called giants'
kettles, appear to favour the view, which at present
is generally adopted by both British and Continental
geologists, that the boulder-clay owes its origin to

land-ice.   I have stated on several occasions that the
view of the marine origin of the boulder-clay agrees
best with the known facts of distribution, and with
the history of the European fauna (pp. 80-86, and
p. 129).   It may be urged that if the lower boulder-
clay were contemporaneous with the British Crags
which succeeded the Red Crag, how can we explain
the fact that these crags contain plenty of shells,
while in the lower continental boulder-clay there
are scarcely any?

But as yet our knowledge of the conditions of life
of the marine mollusca and of their distribution is
extremely scanty.   We are apt to imagine that the
bottom of the sea is covered by a more or less
uniform thick layer of shells; but whenever a careful
survey of the nature of the deposits now forming
there has been made, such is by no means found to
be the case.   Some of the best results obtained by
that useful body, the Liverpool Marine Biological
Committee, have been precisely in this direction.
A most interesting account has been published by
Professor Herdman and Mr. Lomas on the floor
deposits of the Irish Sea, in which the authors state
(p 217), that "a place may be swarming with life
and yet leave no trace of anything capable of being
preserved in the fossil state, whereas in other places,
barren of living things, banks of drifted and dead
shells may be found, and remain as a permanent
deposit on the ocean floor."

Owing to the fact of the peculiar geographical

position of Scandinavia at this time—an isthmus of land with a high mountain range lying between the warm Atlantic and the cold Arctic Sea—the snowfall must have been excessive, and large glaciers were evidently forming. These produced icebergs as soon as the lower parts had advanced to the Baltic coast-land and deposited their detritus in the sea. Immense masses of mud and stones were thus cast to the bottom of the sea, and under these circumstances no delicate mollusca or other marine life probably could have developed within a considerable distance from the shore. To judge from the direction pursued by the majority of the boulders from their source of origin, the prevailing current during the deposition of the lower boulder-clay was from north-west to south-east. It is possible that little marine life, except free-swimming forms, would have been able to live within the Russian area of this sea. But the free-swimming larvæ of molluscs and other surface species were not prevented from passing from the White Sea south-westward, and in sheltered localities where little or no mud deposition was going on, these no doubt might have developed into adults on the sea-floor. It is quite conceivable, therefore, that in one portion of the North European Sea, which was fully exposed to the destructive influences of the iceberg action, the fauna was scanty or totally absent, while in another part there lived a fairly abundant one. The unfossiliferous state of the lower continental boulder-clay does not, therefore, offer any serious difficulty to the supposition that some of the so-called Newer Pliocene Crags of the east coast of England were laid down at the same time by the same sea.

This would also explain how the Arctic species come to inhabit the Caspian, as the old Aralo-Caspian Sea could have had some communication (Fig. 12, p. 156) with the North European Sea. And this again offers an explanation of the otherwise mysterious occurrence of the Caspian *Dreyssensia polymorpha* in the lower continental boulder-clay.

The climatic reasons for the supposition that the boulder-clay is a marine deposit have already been given (p. 66). However, it may be asked what about the glacial flora which has been proved to have existed all over the plains of Northern Europe?—what about the relics of this same flora which still linger on in a few localities to the great delight of the systematic botanist? They have been spoken of as indications of a former Arctic climate in Europe. The presence of an Arctic species such as *Dryas octopetala* in any of the pleistocene deposits is often looked upon as an absolute proof of a very severe climate having prevailed at the time they were laid down. Professor Geikie tells us that the South of England was clothed with an Arctic flora, when the climate became somewhat less severe than it had been during the climax of the glacial cold (p. 398). Relics of such a flora have been detected at Bovey Tracey, in Devonshire, the Arctic plants found comprising

Betula nana and *B. alba, Salix cinerea* and *Arctosta-phylos uva-ursi.*

Now three of these four species of plants are still natives in the British Islands, and all are forms which probably came to us with the Arctic migration which I described in Chapter IV. They travelled south with the reindeer, or before it, and may have covered large tracts of country at the time. With the increased struggle for existence on the arrival of the Siberian and Oriental migrants, they have probably been evicted by these more powerful rivals. A discovery of their remains does not necessarily indicate that a great change of climate has taken place since they lived in the country. And certainly these Arctic plants cannot be taken as indicating a low temperature, for it has been shown that Alpine plants are mostly intolerant of very low temperatures. "Arctic and Alpine species in the Botanical Gardens at Christiania," says Professor Blytt (p. 19), "endure the strongest summer heat without injury, while they are often destroyed when not sufficiently covered during the winter." Similar observations have been made in other countries. For this reason they have to be generally wintered in frames in the Botanic Gardens at Kew and Dublin, and are thus exposed to higher temperatures than at present obtain in the British Islands. This fact suggests that the Alpine and Arctic plants really did not originate in countries with cold temperatures. They probably made their first appearance long before the Glacial period—

perhaps in early Tertiary times—chiefly in the Arctic Regions, which at that time had a mild climate. They have since become adapted to live in cold countries where they flourish, provided they receive sufficient moisture in the summer, and are protected from severe frost in the winter by a covering of snow. When we carefully examine the present range of Arctic plants in the British Islands, a curious fact presents itself which no doubt has frequently been noted by botanists, viz., that some of the most characteristically Arctic species, and some which are often quoted by glacialists in support of their theories, flourish at the present moment in very mild situations. I have already referred to the fact that the Mountain Avens (*Dryas octopetala*) abounds in the west of Ireland (County Galway) down to sea-level. Now it is well known that the mean winter temperature of that part of Ireland resembles that of Southern Europe, being no less than 12° F. (=7° Cent.) above freezing point. The plant, of course, is here a native, and not introduced. This instance shows clearly, that as long as more vigorous competitors are absent, and as long as it is not exposed to severe frost or undue dryness, this and allied species do just as well in a mild climate as in their native Arctic home.

In his interesting essay on the distribution of the Arctic plants in Europe during the Glacial period, Professor Nathorst adduces the fact that all the localities but one, in which remains of such plants have been discovered, lie either within or close to the limits of the maximum extension of the supposed northern ice-sheet, or within those of the former Alpine glaciers. Whether we look upon the boulder-clay as a marine or a terrestrial product, it is quite conceivable that, in many instances, the remains of the Arctic plants may have been carried by ice to great distances from where they grew. The probability, however, is in favour of most of them having lived where their remains are now found. Now, it is a remarkable fact, that the single instance in Europe of a deposit of Arctic plants having been found far removed from the maximum extension of the northern ice-sheet is the one quoted above, viz., at Bovey Tracey, in Devonshire. Even up to recent times Arctic plants may have persisted at Bovey Tracey just as they do in Galway under the influence of a mild coast climate. Similar circumstances may have led to their survival along the shores of the sea which deposited the North European boulder-clay, while they moved northward from the Alps along with the glaciers, which always supplied them with an abundance of moisture. Alpine plants probably became exterminated in the plain of Central Europe at a much earlier period.

SUMMARY OF CHAPTER V.

What has been spoken of in the earlier parts of this book as the eastern migration, refers in a general way to the animals which have come to England from the east. But these are by no means natives of one country alone. We can trace a number of the British mammals to a Siberian origin, and also some birds; among many of the lower vertebrates and invertebrates, however, there are few species which have reached us from Siberia. They may have had their original homes in the Alps, in Eastern Europe, or in Central and Southern Asia, and have joined in their westward course the later, more quickly travelling mammals. Many instances are given from all the more important groups of animals to show how we may proceed in approximately identifying the home of a species.

The periodical invasion into our continent of Pallas's Sand-grouse and other birds, suggests an explanation as to the cause of the great westward migration in former times of the Siberian mammals. Since a considerable amount of fossil evidence is available to show the path of migration pursued by these mammals, other important problems, such as the time of their arrival in Europe and the geographical conditions surrounding them, may perhaps be approximately ascertained, and thus throw much light on the general features of the European fauna. It has been proved by Professor Nehring that the Siberian mammals arrived in Eastern Europe after the deposition of the lower continental boulder-clay. He believes that the climate of Germany at that time had ameliorated so far, after the maximum cold of the Glacial period, that steppes with a Siberian fauna could exist. Other groups, such as the Mollusca, however, do not support Professor Nehring's theory, and in order to arrive at an independent solution of this and the other problems referred to, a short history is given of the Siberian fauna. Recent geological ages have witnessed the arrival in Southern Europe of mammals now almost confined to the arctic and subarctic regions. In Siberia, on the other hand, many southern species penetrated, apparently about the same time, to the extreme northern limits of that country. The greatest authority on the Siberian fossil fauna, Tcherski, believes that this took place in pliocene times, the gradual retreat occupying the whole of the Glacial period. If this were correct, the retreat from the Arctic Regions would have occurred at the same time when, according to our European authorities, Professors Nehring and Geikie, the much more southern parts of our continent were already uninhabitable. But Siberia could not have supported the large mammals at all at a time when Europe was uninhabitable, as it would be difficult to conceive under what geographical conditions the climate of the latter was arctic and that of the former temperate. If the whole fauna was driven into Southern Asia, how is it that the Siberian invasion of Europe occurred immediately after the deposition of the lower boulder-clay, that is to say, after the earlier part of the Glacial period? The difficulty can be met by the supposition that both Europe and Siberia had a temperate climate at that time. This view is supported by certain evidences, fully described, of a connection between the Caspian and the White Sea, which would have had the effect of influencing the climate. The Siberian fauna would thus have been prevented from spreading westward in Pliocene and early Glacial times. But on the disappearance of the marine connection, a way would have been opened into our continent, which again had an effect on the climate. The latter would have become sensibly colder and thus have reduced the habitable area of the Siberian fauna.

Such geographical conditions would have been incompatible with a great northern *mer de glace*, and the boulder-clay in Northern Europe could not have represented a ground moraine but is a marine deposit. The sea is supposed to have covered the Northern Russian and German plains, and into it icebergs discharged the detritus which had accumulated on them when they were still Scandinavian glaciers.

As regards the time of the arrival of the Siberian migrants in Europe, the English Forest-Bed gives us an additional clue to its determination. Since Siberian migrants are unknown from earlier deposits than this, it is reasonable to suppose that they arrived in England about the time when it was laid down. But since they appear in Germany in the inter-glacial beds subsequent to the deposition of the lower boulder-clay, the former are probably contemporaneous with the Forest-Bed. Some of the deposits generally regarded as upper pliocene by British geologists would therefore have to be classed with the lower continental boulder-clay as lower pleistocene. In connection with this theory some interesting faunistic data are given which seem to support it.

In conclusion, the former presence of Arctic plants in Central Europe and their bearing on the climatic problems are discussed.

CHAPTER VI.

THE ORIENTAL MIGRATION.

THE Oriental migration is closely related to the Siberian. Both have originated within the Asiatic continent, and in many respects a strict line cannot be drawn between them. There can be no doubt that some of the species which we regard as Siberian migrants had their original home in more southern latitudes, and thus may have formed part of the older Oriental migration. The home of that migration I take to be Central and Southern Asia, that is to say, everything south of the Altai Mountains and the Caucasus. Its members have reached Europe across an old land-connection which united Turkey, Greece, and Syria, while the Siberian animals invaded our continent to the north of the Caspian and Caucasus.

The Siberian immigrants into Europe on the whole are not very numerous, but it is different with those from the more southern parts of the Asiatic continent. The members of the Oriental migration form a very large percentage of the European fauna. No other migration has affected our continent so powerfully, because it continued uninterruptedly for

245

a very long time. Hence its results can be traced from one corner of Europe to the other. We have seen that the Siberian migration only commenced after the first portion of the Glacial period had passed away. The Oriental, however, persisted throughout, or at any rate for the greater part of that period. It commenced ages before it, in miocene times, or even earlier. And as the Ægean Sea, which broke up the highway of the Oriental migrants, is only of recent formation, there was a steady westward march for a very considerable time. No doubt the migration was also favoured by the fact that scarcely any formidable barriers had to be crossed.

Many instances might be quoted of the same species forming part of the Oriental and also of the Siberian migration, but as a rule the Siberian migrant belongs to a distinct variety, or has such well-marked racial characters as to be at once detected from its more southern relative. Among the examples of Oriental migrants which I have occasion to bring forward, such instances will be specially dealt with.

In its wild state the Red Deer (*Cervus elaphus*) is almost extinct in the British Islands, though it still occurs in the moorlands of Devonshire and Somersetshire in England, in the south-west of Ireland, and in some localities in Scotland. Fifty years ago it was also found wild in several other of the Irish western counties; and in the seventeenth century it was common in most of the mountainous districts of

Ireland. Its remains have been found fossil in the
marls and caves of Ireland, and in the Forest-Bed,
as well as in a large number of caves in England.
The history of the Red Deer in other countries
is very similar. In Scandinavia it flourished as
far north as the sixty-eighth degree of latitude,
whereas it is now quite extinct on the mainland, though still lingering on in some of the
western islands. Denmark and Switzerland know
it no more, and it is almost extinct in Belgium.
Nearly throughout Europe where it occurs, its
numbers are diminishing, greatly owing, perhaps, to
the relentless persecution by man, but its gradual
disappearance must likewise be partly due to other
causes. Formerly it inhabited every country of
Europe and all the larger islands. It still exists
in Corsica and Sardinia, and at an earlier period
it was also met with on the island of Malta.
The Red Deer found in Corsica and Sardinia is
smaller than that inhabiting Central Europe, and is
by some authorities regarded as a distinct species,
which has been named *Cervus corsicanus*. But
Sir Victor Brooke has pointed out that the antlers
of some of the Scotch Deer agree in every point
with those of the Sardinian species. Indeed, the
West European Red Deer altogether is a small-
antlered form, compared with the Eastern one. This
character, however, is only a racial one, and not of
specific value. In the pleistocene deposits of Eastern
and Central Europe, a very large-antlered race has

been discovered, and identified by Professor Nehring with *Cervus canadensis*—the Canadian Red Deer. Tcherski, the Siberian traveller, believed that *Cervus canadensis* was identical with, or a variety of, the Asiatic species of Deer, *Cervus eustephanus*, *Cervus xanthopygus*, and *Cervus maral.* Some authorities —and to these belong Mr. Lydekker—think that we ought perhaps to regard the whole number of Red Deer-like forms as local varieties of one widely-spread species. Besides the deer already referred to, the following belong to this same group:—*Cervus cashmirianus*, *Cervus affinis*, *Cervus Roosvelti*, from North America, and the North African *Cervus barbarus*.

The question now is, where have these varieties originated? Or, if we go to the root of the matter, where is the original home of their ancestors? Considering that so many *Cervidæ* have been found in French and English pliocene deposits, and that remains of the Red Deer occur not only in the English Forest-Bed, but have been found associated with those of the Pigmy Hippopotamus in Malta, it would only be reasonable to suppose that the genus *Cervus* had originated in Europe. It might also be argued with equal force that the Red Deer had its birthplace in our continent. But when we carefully study its present range this verdict cannot be accepted. The view of the Asiatic origin of the Red Deer, so ably maintained by Köppen, corresponds far better with its present distribution, especially if we look upon the Asiatic, North American, and North African forms as varieties of the same species.

If the Red Deer were of European origin, it must have come into existence at a time when Malta was part of the mainland, when North Africa and the British Islands were connected with the continent of Europe, and of course before the deposition of the Forest-Bed. Such land-connections existed probably during the Pliocene Epoch. Migrants would have wandered from Europe into Asia. These would have developed into larger races, which again furnished emigrants for North America. The latter crossed by the old land-connection which once joined America and Asia at Behring's Straits. During pleistocene times the large Siberian race would now have re-migrated to the home of its ancestors in Europe, for we find the remains only in Central and Eastern Europe, indicating that an invasion of the Red Deer from Asia must then have taken place.

Against this view of the European origin of the Red Deer, it may be urged that deer are known from Indian as well as from European pliocene deposits, and that a migration could have taken place from the Oriental Region to Europe just as easily as from the latter to Asia. The majority of the species of the genus *Cervus* (in a wide sense), moreover, are Asiatic, ranging to Borneo, Sumatra, and the Philippine Islands, all of which islands have been separated from the mainland for a considerable time. Finally, the original home of a species, as we have learned, generally corresponds with the centre of its geographical range, and this lies in the case of the Red Deer in Central Asia.

One of the highest authorities on the deer family, Sir Victor Brooke, also was of opinion that the *Cervidæ* originated in Asia, and from there spread east and west. Of the two divisions into which true deer are divided, viz., the *Plesiometacarpalia* and the *Telemetacarpalia*, the former is almost confined to the Old and the latter to the New World. The only North American species belonging to the first division is the Canadian Red Deer, which fact clearly indicates its recent immigration to that continent.

There were probably two distinct migrations of the Red Deer into Europe. An older one coming from Asia Minor into Greece, which stocked Sardinia, Corsica, Malta, and North Africa in the first place, when these were still connected with one another. This same migration likewise affected western continental Europe, the Irish Red Deer being probably the descendant of this very ancient stock. The latter entered the island when it was still part of the Continent. The later migration of a larger form came from Siberia and spread mainly over Eastern and Central Europe, but it appears that it also reached England, although there is no evidence of any of these Siberian deer having ever inhabited Ireland.

The range of this deer, therefore, to some extent corresponds to that of another described on p. 153. We found then that two races of Reindeer had migrated to the British Islands—one from the Arctic Regions, and the other from Siberia, but that only the former had reached Ireland.

The so-called Irish Elk (*Cervus giganteus*) has been referred to the Oriental migration, but, as stated below, it has some claims to be regarded as a European. Unfortunately it is now extinct; it seems not unlikely, however, that it inhabited Ireland when man had already made his appearance on the island. Although its remains are found in such extraordinary abundance in Ireland, it certainly did not originate there. It lived also in England and Scotland, and in the Isle of Man, in France, Denmark, Germany, Austria, North Italy, and Russia. Its remains have been discovered even in Siberia. It must either have originated in Europe and then migrated to Asia, or have had its birthplace in Asia and wandered to Europe. There is nothing to lead any one to assert positively that either of these two continents was the one in which the original home of the Irish Elk was situated, and we can only be guided in this case by the history of its nearest relatives. These are the Fallow Deer (*Cervus dama*). There are two very closely allied species, the Persian and the European, but several others have been discovered in the Forest-Bed and the pliocene deposits of the Auvergne. As no remains of the Fallow Deer are known from Asia, it seems probable that it and also the Irish Elk originated in Southern Europe, and only invaded Asia in early pleistocene times.

The Mammoth (*Elephas primigenius*) is a familiar example among a large number of mammals which have come to us about the same time from Asia by the Asia Minor route. It had a much wider range than the Irish Elk, since its remains have been discovered in a large number of European localities as far west as Ireland, also in Siberia, and even North America. Though we have had *Proboscidea* in Europe from the Middle Miocene onwards, Mr. Lydekker (*d*, p. viii.) holds that "our comparatively full knowledge of Lower Miocene and Upper Eocene mammalian faunas of the greater part of Europe and North America, renders it almost certain that neither of those regions was the home of the direct ancestors of the *Elephantidæ;* and we must therefore look forward to the discovery of mammaliferous Lower Miocene or Upper Eocene strata in some other region of the (probably old) world which may yield these missing forms."

The genus *Elephas* makes its first appearance in the Upper Miocene of. India.   Our European *E. antiquus* is, according to Professor Zittel, probably identical with *E. armeniacus* of Asia Minor, while *E. meridionalis* agrees in all essential characters with the Indian *E. hysudricus.* The Indian and European species of fossil elephants altogether are

very closely related, and the supposition that they
all have had their original home in the Oriental Region
offers, I think, no serious obstacle. The view of the
European origin of the mammoth especially is open
to very serious objections. It does not occur in any
European pliocene deposits, and could not therefore
have originated in our Continent until pleistocene
times. That it should then have commenced its
travels through Europe and Siberia to the New
Siberian Islands and North America seems almost
an impossibility. But if we suppose the mammoth
to have had its home in India in pliocene times, it
could then easily have migrated to all the parts of
the world where its remains have been discovered.

Of the Asiatic mammals still living, some have
only just crossed the borders of Europe and then
died out again. Similar cases have been referred to
in discussing the Siberian migration. Thus remains
of the camel have been found in Roumania and in
Southern Russia in pleistocene deposits. Others have
lingered on to the present day. *Crocidura etrusca*,
for instance, still lives in Southern France, Italy,
Sicily, and North-western Africa. All its nearest
relations are typically Oriental species. In spite of
the fact that a *Crocidura* is known from French and
German miocene deposits, the general range of the
genus suggests an Oriental origin. In early Tertiary
times a section spread into African territory and
another eastward as far as the island of Timor. This
may possibly have happened in miocene times, when

a few species likewise found their way into Europe. Many other mammals have wandered still farther west, and now form an important percentage of the European fauna.

Of Birds, too, a large number might be mentioned which had their home in Asia and have found their way to Europe with the Oriental migrants. A few instances have already been alluded to, and some additional ones may be specified. at random, without attempting to give a complete list.

Some of the Wagtails (*Motacilla*), as I mentioned in the last chapter, have certainly come to us with the Siberian migration; but others seem to be Oriental, such as *Motacilla melanope*, which is resident in Southern Europe and migratory in the North. *M. campestris*—the Yellow Wagtail—has a most peculiar discontinuous range. One colony breeds in the British Isles and Western Europe generally, where it is known as a summer visitor, retiring to West Africa during winter; another is found from South-east Russia to Turkestan in summer, and winters in Southern Africa. This fact may possibly be due to two distinct migrations from Asia having taken place : an earlier one from the South-east—that is to say, an Oriental one—and a Siberian one more recently. In this case the members of the two migrations have not become sufficiently differentiated to be regarded as distinct varieties. Though most of the Wagtails have a somewhat northern range, none (except perhaps *M.*

borealis) are truly Arctic; and indeed, as almost all of
them pass the winter in southern latitudes, it may
be* assumed that they are of southern and not of
northern origin.

The Dippers (*Cinclus*) are practically unknown in
the Central European plain, but they occur in
Western Europe as far north as Scandinavia, also
in the Alps, Carpathians, and Southern Europe, including Sicily and Sardinia. Some authorities distinguish three species, others only one. As a matter
of fact, the difference between the three forms is very
slight, and their nests and eggs are undistinguishable.
Eight other species have been recognised, and all
these are either Asiatic or American. As one of the
American forms is peculiar to Peru and another to
Ecuador and Columbia, and since the genus as a
whole is a mountain-genus, it probably is an ancient
one. Its European range alone, however, implies that
it has inhabited our continent for a considerable
time and is no new-comer. We may look upon it as
of Asiatic origin. The ancestors have spread east
and west, the European species having arrived with
the earlier Oriental migrants, and wandered along
the Mediterranean at a time when the geographical
conditions of that sea were vastly different from what
they are to-day.

Not quite so ancient as the Dippers, but likewise Asiatic in their origin, are the Bullfinches
(*Pyrrhula*). The closely allied Pine-Grosbeak (*Pinicola enucleator*) has already been referred to (p. 191)

as a member of the Siberian migration. The
distribution of the European Bullfinch (*P. europœa*)
is very interesting, as it occurs in two distinct
forms, by some authorities regarded as races, by
others as species. In all probability these two
races owe their origin to two different migrations
from the same ancestral stock. We may suppose
that *P. europœa* came to Europe along with the
Oriental migration, spreading chiefly over the south
and west, while another branch developed in Siberia
into the larger and more brilliant race (*P. major*),
which subsequently entered the neighbouring continent with the Siberian fauna. The latter race
inhabits, according to Mr Saunders, Northern and
Eastern Europe, and also Siberia. All the other
species—there are eight more—except one, are found
in Asia. This one species, which inhabits the Azores,
appears to be more closely related to one of the
Siberian bullfinches than to the European. It stands
isolated, and is an extraordinary instance of discontinuous distribution, as no Bullfinch inhabits either
Madeira or the Canary Islands. We must assume
that the form connecting it with the Asiatic probably lived in Southern Europe, and has become
extinct.

One of the most typically Oriental genera of birds
is *Phasianus*, to which our Common Pheasant belongs.
Out of twenty species, nineteen are found exclusively
in Asia, most of them being confined to the central
plateaux of that continent. Only one species passes

the confines of Asia into Greece, Turkey, and Southern Russia. This is *Phasianus colchicus*. Formerly, however, the Pheasant appears to have had a wider range in Europe, for three species are known fossil from France. Altogether, it is not quite certain whether the Pheasant is not really an indigenous bird in the British Islands, having survived from preglacial times. It is believed that the Romans brought it to England, but there is no record of an introduction at that time.

Among the older Oriental bird migrants might be mentioned the Fire-crested Wren(*Regulus ignicapillus*), which has even occasionally visited England. It becomes commoner as we go south-eastward. In Asia Minor it is more abundant than the Gold-crest; and throughout the year it is resident in Southern Europe, where it occurs in Turkey, Greece, Italy, Spain, Sardinia, and Malta. On the opposite shore, in North-west Africa, it again makes its appearance, and its range extends westward to the Canaries (*R. teneriffæ*) and Madeira (*R. maderensis*).

The genus to which our common Goldfinch belongs, viz., *Carduelis*, is also probably of Oriental origin, and may be looked upon as one of the earlier migrants. That species (*C. elegans*) breeds throughout Europe, except in the extreme north, but it is especially abundant in Southern Europe and North-west Africa. It is also resident in Madeira and the Canaries. Eastward its range extends to Persia. A larger race (*C. major*) inhabits Western Siberia and crosses

17

the European border into Russia. It interbreeds in Siberia with *C. caniceps*, an East Siberian form.

A few instances of Reptiles and Amphibia with a similar range will show that the Oriental migration was not confined to the higher vertebrates.

Two species of the genus *Eremias* (*Podarcis*) occur in South-eastern Europe. This is a genus of Lizards with rather a wide distribution, ranging from Central Asia to South Africa southward and China eastward. Altogether there are twenty-four species, two of which just enter Europe; and of the rest half are Asiatic and half African. Even if the genus were of African origin, it is extremely unlikely that the Asiatic species came by way of Europe. We may assume, therefore, with a fair degree of probability that the two European species wandered westward along with the Oriental migrants.

The genus *Ablepharus* belongs to a family of Lizards in which the legs are either very fully developed, or quite absent as in the Slow-worm (*Anguis fragilis*). It is an ancient genus, having a wide range from Central Asia to Australia on the one hand, and to South Africa on the other. One species of this Scink-like Lizard, viz., *Ablepharus pannonicus*, enters Europe in the south-east, inhabiting Greece as far north as Southern Hungary. In Asia it is found in Syria and North Arabia. This clearly signifies that the Lizard is an Oriental migrant.

Among the Snakes which participated in the

Oriental migration might be mentioned *Eryx jaculus*, whose home is probably in Western Asia. It is known in Europe from the Greek islands of Tinos and Naxos, from Turkey and Southern Russia. Another, a peculiar worm-like form, lives underground in damp earth and under stones—*Typhlops lumbricalis.* This species inhabits the mainland of Greece as well as the Greek islands, and Asia Minor as far as the Caucasus.

A most interesting case of distribution is that of the pretty little Toad so well known on the Continent under the name of "fire-toad" (*Bombinator igneus*). Though some authorities, such as Boulenger, recognise only one form of *Bombinator*,[1] others are of opinion that two well-marked varieties exist in Europe. These are looked upon by Dr. von Bedriaga as good species, but he acknowledges that they are rather critical and difficult to identify. No other species of *Bombinator* occur in Europe. *Bombinator pachypus*, the western race,—or if we choose to call it species,—occurs in France, Germany, Switzerland, Austria, Sicily, and Greece. *B. igneus*—the eastern race—is found in Southern Sweden, Denmark, Germany, Austria, and Russia. The latter has therefore a more northerly and easterly range. The species is not known from Siberia, but makes its appearance again in China in a form which, according to Dr. von

---

[1] Since writing the above account, Mr. Boulenger, in his new work on the Batrachia of Europe, has accepted the specific distinctions between the two fire-toads.

Bedriaga, does not quite agree with either of the two European races.

Now if we supposed *Bombinator* to have originated in Europe, its absence from the British Islands, most of the Mediterranean islands, and the greater part of Scandinavia would not be easy of explanation, while as an Asiatic migrant the European range is more readily understood. Its apparent absence from Western Asia might quite likely be due to the fact that the zoology of that part of the Continent is only now being investigated. The latter has, moreover, undergone great physical changes in recent geological times. The supposition that one migration of *Bombinator* from the south-east has taken place, and then another from the east, seems to explain this case of distribution, as other similar ones, in a most satisfactory manner.

The Tree-Frog (*Hyla arborea*) must be an ancient species, but it is not of European origin. Few genera of Amphibia have a wider distribution than *Hyla*. There are only three species in Asia, Europe, and Africa, the remaining 129 being confined to America and Australia. Two of the three Old World Tree-frogs are so closely allied that until recently they were regarded as mere varieties of one another. These are *Hyla arborea* and *H. chinensis*. The former is found in Asia Minor, Persia, China and Japan, and in most of the Mediterranean islands and Southern Europe generally. It does not occur in the British Islands, Norway, or North Russia, but in

South Sweden, Germany, France, and Spain. It is
also known from North Africa and from Madeira, the
Canaries, and the Salvages. The occurrence of the
Tree-Frog on so many of the Mediterranean islands
is of particular interest, especially as four well-marked
varieties have been distinguished by our leading her-
petologists, so that the more minute features of the
various forms can be traced from island to island,
adding one more proof—if proof were needed—of
their former continuity. Of course, that *Hyla arborea*
must be considered an Oriental migrant seems so
evident that it scarcely needs further comment.

A number of mollusca might be mentioned whose
range indicates that they have migrated to Europe
from Asia Minor. *Buliminus pupa* is one of these.
It is known from Asia Minor, Greece, South Italy,
Sicily, and Algeria. *Buliminus detritus* is perhaps
better known, being common in some parts of
Germany. From there its range spreads east as
far as Asia Minor. Many closely allied species
inhabit Western Asia, to which they are confined,
while others enter on European territory in some
of the Greek islands. *B. fasciolatus* occurs on the
islands of Crete, Rhodes, Cyprus, and in Greece
and Syria. Most of the species of *Buliminus* have
a very restricted range, but *Buliminus obscurus* is
found almost all over Europe, from Ireland in the
west to the Crimea and Transcaucasia in the east.

Whether the sub-genus *Pomatia* of the genus
Helix—to which the so-called Roman Snail belongs—

is of Asiatic origin, or whether some of the species have migrated from Europe to Asia, I am not prepared to say; but there can be no doubt that *Helix pomatia* has reached Western Europe from the east.

On the whole, the number of mollusca which we might point to as having migrated to Europe is not large, the great majority being indigenous to our continent. However, some of the other groups of invertebrates differ very materially in that respect from the mollusca. I cannot leave the consideration of the mollusca without referring to the fact that there appears to be a very important centre of distribution in South-eastern Europe. It is from this centre that many species have spread north and south, east and west. Take, for example, the genus *Clausilia*, a small land-shell shaped like a pointed round tower, and abundant on old walls and tree trunks. In England we have four species of *Clausilia*, in Ireland only two. In the greater part of Spain only our common *Cl. bidentata* occurs. As we go east the number of species rapidly increases. A maximum is reached in South-eastern Europe, where hundreds of different kinds are found. Towards Northern Europe a similar decrease of species takes place. So far the history of the *Clausiliæ* seems perfectly simple. An active centre of origin appears to exist in South-eastern Europe, from which the species radiate out in all directions. But when we come to look more closely into the extra-European distribution of the genus, and especially when we examine its past history, we find that its origin is extremely complex, and dates back to a much more remote period than would have been imagined, had we merely taken into account its present range in our own continent. Professor Boettger, who is the highest authority on *Clausilia*, tells us that the genus is known from the earliest deposits of the Tertiary Era. About 700 species are now known, and these have been sub-divided by Professor Boettger and others into a number of sub-genera. Some of these are extinct, but the great majority are still living. The sub-genus *Phædusa* occurs in the eocene and oligocene of Southern Europe, but it is extinct as far as our continent is concerned. Close upon a hundred species, however, still inhabit India, the Malayan Islands, China, Ceylon, and Japan. Then again, the sub-genus *Laminifera* occurs in the oligocene and miocene of Central Europe, and survives in a single species, *Cl. Pauli*, in South-western France. The groups *Garnieria* of China, *Macroptychia* of East Africa, *Boettgeria* of Madeira, and *Nenia* of South America, have no fossil representatives. We have here some very remarkable cases of discontinuous distribution which testify to the antiquity of the genus, and this is certainly confirmed by the fossil evidence. However, it is hardly likely that the headquarters, as it were, of *Clausilia* have always been in South-eastern Europe. Most of that part of the Continent has been submerged since eocene times more than once. The peculiar distribution of the genus might be explained, I think, if we supposed the original home of *Clausilia* to have been in Southern Asia, that from this centre Southern Europe was colonised, where a new centre developed in oligocene and miocene times, sending colonies off to Madeira and across the old land-connection which united Northern Africa and South America about that time. The most active centre of development then gradually shifted eastward again, while the older centres were perhaps sub-merged during the physical changes in the distri-bution of land and water.

I should have mentioned that the species wandering westward and northward from this South-European centre of distribution, would naturally have joined the migrants which came from beyond the borders of our continent. They might thus appear to be true Oriental migrants, and on a previous occasion I grouped all these together under the term of "Southern Fauna," as I assumed the observer to be stationed in the British Islands. All new-comers from the south-east, south, or south-west of Europe would be to him southerners quite irrespective of their original home, which might be in Southern Europe, Asia, or Africa.

The Swallow-tail is well known to all collectors of Butterflies in England, though it has of late years become very rare and is now confined to a

few localities in the east of England. The members
of the family *Papilionidæ*, to which it belongs, are
mostly large and striking species, and their distribution is therefore more accurately known than
that of the smaller and less conspicuous butterflies.
Only four different kinds of Swallow-tail Butterflies
inhabit Europe, but in Southern Asia and the Malay
peninsula they attain their maximum as regards
numbers; and there we find a great many species of
this genus *Papilio*. Of the four European species only
one, viz., *Papilio hospiton*, is peculiar to Europe; all
the others range into Asia. It would seem, therefore,
as if this genus was an Asiatic one and had migrated
to Europe, and that the route taken was the one from
Asia Minor across to Greece. We have a similar
case in the closely allied genus *Thais* two of the
three European species living also in Asia Minor.
Thais cerisyi inhabits some of the Greek islands, as
well as the mainland of Turkey and Greece.

Another genus of the great family *Papilionidæ*
with which most lepidopterists are well acquainted
is *Parnassius*. What butterfly-hunter has been in
Switzerland without hearing of, or seeing, the famous
Parnassius Apollo? We have four European species of
Parnassius, only one of which is peculiar to our continent, but the locality where it occurs, the Caucasus,
is on the borders of Asia. Almost all the other
species are Asiatic, none however range to the south.
Its headquarters, and I think its original home, are
the mountains of Central Asia. From there it

has spread—some species to the Himalayas, and a few to Europe and North America. But these migrations are not of very recent date. *Parnassius* no doubt arrived accompanied by a large number of other Central Asiatic mountain insects and plants. I shall refer to the latter again when dealing with the origin of the Alpine fauna, but meanwhile it might be mentioned that the famous Swiss "Edelweiss" (*Leontopodium alpinum*), which we are accustomed to regard as a typical Alpine plant, is certainly of Asiatic origin. In some parts of Southern Siberia it is one of the common meadow-flowers, and ranges from there south into Kashmere, but not northward. Like the *Apollo*, it does not occur in Scandinavia or Northern Siberia. Both plant and insect evidently migrated from Central Asia, directly westward along the southern border of the sea, which extended from that region as far as the European Alps in early Tertiary times. At that time the Caucasus was possibly still connected with the Balkan Mountains, across what is now the Black Sea, and that may have been the highway on which they travelled west.

Some of the Clouded-Yellows—butterflies appertaining to the genus *Colias*—formed part of the Oriental migration. The genus is undoubtedly of Asiatic origin, and while many of the species have turned northward, ranging across Siberia and North America, others have taken a southern and westward turn and thus reached Europe. We have two

Clouded-Yellows in Western Europe, and both of them must have come with this migration.

' A very good example of an Oriental migrant is *Daňais chrysippus*, a magnificent butterfly found in Greece and Southern Italy. In Asia it is known from Syria, Persia, and from the whole of the southern portion of the Continent. The genus *Danais* (in its wide sense) is a large one, and principally occurs in the warmer regions of Asia. Three species are found in North America and only one in Europe.

Among the beetles belonging to this migration, there is one of very considerable interest from a distributional point of view, for all the species of the genus—even the whole family to which the genus belongs—are what is known by zoologists as "Commensalists." These are animals habitually associating and living in close connection with others with which they are not tied by any family relations or kinship. Such a state of close and permanent friendship is called "commensalism." Now it appears as if the members of this family of beetles (*Clavigeridæ*) had of their own free will formed such a close connection with colonies of ants—sometimes with one species, sometimes another. They are the permanent guests of the ants, and in return they secrete a fluid which is apparently highly prized by them. All of the *Clavigers* are provided with peculiar club-shaped antennæ, with which they ungraciously beat their hosts, when they are in want of food. According to some authorities, they even occasionally gnaw at the pupæ and larvæ of the ant with which they live.

Such beetles naturally can only have extremely limited means of distribution, and they are comparable in that respect with the woodlice of the genus *Platyarthrus*, to which I have already had occasion to refer. All the species of *Claviger* are confined to Europe, chiefly to the south, but one species, *Cl. testaceus*, has wandered farther north and occurs in the nest of the ant *Lasius flavus* in the south of England, Ireland, and Scotland. Though none of the *Clavigers* can be claimed as Oriental migrants, the centre of distribution of the genera belonging to the *Clavigeridæ* is in Southern Asia, and it is probable that the ancestors of the European *Clavigers* have spread westward from that region to Europe, eastward to Australia and Japan, and southward to Madagascar and South Africa. The genus *Hopatroides*, belonging to the same family as the so-called Spanish-fly (*Tenebrionidæ*), has twelve species in Western Asia and Greece. One only, *H. thoracicus*—an instance of discontinuous distribution—occurs in Andalusia. *Amphicoma* is represented in Western Asia and the Balkan peninsula by fifteen species, while three others are met with in North-west Africa and Southern Spain.

A genus of Dragon-fly, *Onychogomphus*, has in Europe a somewhat similar distribution to *Claviger*, but it has besides a very extensive foreign range. There are altogether thirty-five species; of these ten are Holarctic, twelve Oriental, five Mascarene, and

eight Ethiopian. The centre of distribution is therefore in the Oriental region, and we may assume that
in all probability the genus has originated there,
the European species having travelled west with
the Oriental migration at an early date of the
Tertiary Era.

Ryothemis, another genus of Dragon-flies, has
originated perhaps somewhat farther east than the
last, for no less than thirteen species are found
in Australia, a like number in India, five in Madagascar and Africa, and five in the Holarctic region.
Both of these genera are entirely absent from
America, and they have possibly travelled to Europe
together.

Among the European *Orthoptera*—the group to
which our Earwigs and Grasshoppers belong—there
are also a good many instances of Oriental migrants.
One of the most striking of these is the curious
"praying insect" (*Mantis religiosa*). It occurs all
over Southern Europe, and ranges as far north as the
north of France. It is also found in Southern Germany and in Austria, and has a vast extra-European
range. There are even records of its occurrence from
all parts of Southern Asia and Java and a great part
of Africa. That it belongs to an extremely ancient
genus is testified by the fact of its presence in
Mauritius, Japan, Australia, New Zealand, South
America, and Madagascar. The genus *Bacillus*—to
which the typical Stick-insects belong—has a somewhat similar geographical distribution. But no less

than four species of *Bacillus* are known from Europe,
according to our great authority Mr. Brunner von
Wattenwyl—all from the south; and some of these
also range into North Africa. There are thirty-two
other species distributed over Southern Asia, Africa,
Australia, New Zealand, and the Sandwich Islands.

Volumes, indeed, might be filled with lists of
species and genera of terrestrial invertebrates of
Oriental origin, but I will not weary the reader
with further enumeration of such instances. Just
two more, however, before concluding, as I have
not alluded to the large group of the *Arachnida*.

Two peculiar spider-like genera, viz., *Galeodes* and
Rhax, are found in Southern Europe. Both occur
also in North Africa, and in Western and a portion of
Southern Asia. As the whole family altogether has
an Asiatic character, I cannot agree with Mr. Pocock,
who considers them of European origin and believes
that they are migrating eastward.

But not only terrestrial forms migrated to Europe
from Western and Southern Asia. Freshwater
species also took part in this great Oriental migration. I need only refer to the freshwater Crab
(*Thelphusa fluviatilis*), with which Southern Euro-
peans are familiar. It is the sole representative of a
large genus which ranges east as far as Australia and
southward to Madagascar and the Cape of Good
Hope. The European species is found in Turkey,
Cyprus, Greece, Southern Italy, Sicily, North Africa,
Southern Spain, Syria, and Persia.

There is a corresponding flora with a range exactly
similar to that of some of the animals quoted. Thus
the Balkan Rhododendron (*Rhododendron ponticum*)
is again met with in the western Mediterranean
region in Southern Spain. The Cedar occurs in
local varieties in the Himalayan Mountains, in the
Lebanon, and the Atlas Mountains. Both of these
are instances of discontinuous distribution, a proof of
their antiquity; but a large number of plants have a
continuous range between Asia Minor and Spain.

On looking through these few instances of what have
been called Oriental migrants, one cannot help being
struck by the fact that the species after their entry
into Europe evidently did not all follow the same path
during their westward advance. We have seen that
a good many seem to have travelled either due west
or north-west on entering our continent from Asia
Minor. They may now perhaps be found in Greece,
Southern Italy, Algiers, and Spain, also probably on
some of the intervening islands in the Greek Archipelago, in Sicily, Sardinia, and Corsica, or they may have
travelled north-east and occur in the Alps. This distribution indicates undoubtedly, as I have already set forth
in another memoir (*c*, p. 459), that land extended from
Asia Minor across Greece to Southern Italy, that the
latter again was disconnected with Central Italy, but
united with Sicily, Sardinia, and Tunis, and that the
Straits of Gibraltar did not exist at the time when
these species migrated westward. Some species
are only to be found as far west as Southern Italy,

while others occur in Central and Northern Europe,
scarcely in the South, and not at all in the
larger Mediterranean islands or in North Africa.
This appears to me to indicate that the late comers
from the east found that geographical changes had
taken place in Southern Europe which prevented
them from following the same track as the older
immigrants. They were now obliged to turn directly
northward and then westward. It may be asked,
why should not the earlier migrants have taken the
same route? This question will be answered immediately. Meanwhile it should be clearly understood
that there probably was an older and a newer migration from the east. The Oriental genera—from
whose general range we know that they must be
very ancient indeed, such as *Mantis* and *Bacillus*—
are almost invariably confined to Southern Europe.
There they are frequently found on some of the
Mediterranean islands. The earlier migrants therefore went westward and the later ones northward.

Let us now inquire a little into the reasons why
such different courses were pursued by the migrants
—why the Oriental migration divided into two
streams, an older and a newer.

During early Tertiary times, and probably throughout the Miocene and Pliocene Epochs, the Ægean
Sea did not exist. From the island of Crete to the
Peloponnesus, and from Asia Minor to Thessaly
and Macedonia, stretched a vast and fertile plain

dotted over with numerous freshwater.lakes. Gradually the sea encroached upon this land from the
south, owing chiefly to extensive subsidences having
taken' place. Only very recently, says Professor
Suess, did the whole of the Ægean continent subside
(i., p. 437). Huge cliffs of levantine freshwater
deposits now mark the new coast-line, and the Mediterranean advances steadily towards the Black Sea
and the Sea of Asov. A new order of things is now
established, continues the famous author of *Das
Antlitz der Erde;* where there were high mountains
we now behold a deep sea, in some places many
thousand feet deep. All this took place quite
recently,—geologically speaking,—certainly in post-
glacial times; and man may even have witnessed
these imposing events. Most geologists admit the
correctness of these views. They are, moreover,
built upon such solid geological evidence, that even
if the science of zoogeography had not yet taught us
anything, naturalists would not hesitate in accepting
them.

Animals and plants were free to migrate from
Central and Southern Asia to Greece by land for
untold ages. The vast accumulation of mammalian
bones which have been discovered at Pikermi, and so
ably described by Gaudry, are probably to a large
extent the remains of Asiatic immigrants to Europe.
Many of these resemble forms still living in South
Africa, which implies that a highway existed also at
that time between Asia and Africa. Among these is

18

a giraffe and antelopes closely allied to African species, and other most interesting mammals.

In still earlier European deposits—the Miocene—we find the ancestors of modern Elephants, which are probably of Asiatic origin. The remains of several kinds of monkeys occur, whose nearest relations are now confined to Southern Asia. Altogether the fauna bears a strong Asiatic facies. Many of our European terrestrial invertebrates probably arrived about this time from Asia. The struggle for existence being keener and the facility for migration much greater in the higher vertebrates, they—or at any rate the mammalian faunas—were subjected to more rapid changes than the invertebrates. I have repeatedly expressed my belief that a great number of our familiar insects and mollusca inhabited Europe long before our present mammals came into existence.[1]

Let us now follow one of the miocene Oriental migrants starting from Central Asia on its way to Europe. Very soon after leaving its home, it must have encountered a sea which extended at that time from the Eastern Mediterranean to the borders of Afghanistan. In following a westward course, the emigrant was compelled to keep along the northern shore of it. We do not know the state of the physical geography of the region between the Black

---

[1] In some cases the accuracy of this view is proved by fossil evidence, *Helix rotundata*, a common and widely spread British species, having been found in miocene strata near Bordeaux.

Sea and the Tianshan Mountains, but it seems certain that a considerable extent of dry land enabled a wanderer from Central or Southern Asia to reach the Balkan peninsula by skirting the northern shore of that large miocene sea. No miocene deposits occur north of Teheran or of the Upper Euphrates, nor are they known from the islands of the Ægean Sea or the lands surrounding it. From the Balkan peninsula it was possible for our migrant to reach the European Alps, which were then slowly rising as a peninsula out of the western portion of the great miocene sea. What are now the Alps was then hilly ground, which was being raised from the bottom of the sea. It was no doubt connected with the Balkan peninsula, so that an intercourse of species could take place between this newly-formed peninsula and Central Asia. I say peninsula, because the miocene sea almost completely surrounded it. From the Western Mediterranean a wide gulf extended up the Rhone valley into that of the Rhine as far north as Maintz. Then skirting along the northern outliers of the Tyrol, the gulf can be followed as far east as Transylvania. It is quite probable that it extended much farther east still, but there is as yet no geological evidence forthcoming. At any rate, our Asiatic migrant turning northward from the Balkan peninsula found its farther progress barred once more by an arm of the same sea which in its earlier peregrinations had stopped it from going south (cf. Suess, i., p. 406).

In later miocene times the sea does not seem to
have surrounded the Alps to the same extent as it
did before, but it certainly extended from the Eastern
Alps to the shores of the Sea of Asov, so that the
direct northward passage was still more or less barred
to the Oriental immigrants. At the same time Alpine
species were now able to emigrate to the North
European provinces. During the last stages of this
epoch, the same sea increased its area very considerably in an eastward direction. One continuous
expanse of water now stretched from the Alps as
far as the Sea of Aral in Central Asia, perhaps even
farther.

During pliocene times especially, the northern parts
of the Balkan peninsula were occupied by a series of
freshwater lakes, while Greece was joined to Southern
Italy, Sicily, and Tunis. Central and Northern Italy
were represented by a long narrow peninsula connected in the north with the Alps. Corsica and
Sardinia were joined to Sicily, and the Straits of
Gibraltar did not exist. When I first published my
views regarding these geographical conditions of the
Mediterranean area, Professor Depéret was good
enough to send me his criticisms from a purely
geological standpoint. He is of opinion that though
Sicily and Sardinia might at this time have still been
connected with Tunis, the Straits of Messina must
already have been formed—in other words, Southern
Italy and Sicily could no longer have been connected
with one another. This opinion is based upon the

fact that in the upper strata of the enormously thick
Sicilian pliocene deposits are found a number of
arctic or subarctic species of mollusca which are
entirely foreign to the Mediterranean fauna. It is
generally supposed that these reached the Mediterranean area by the newly opened Straits of Gibraltar
in later pliocene times, and that the lower Sicilian
deposits must therefore have been laid down earlier.
So far the deductions are perfectly correct, if we assume
the northern mollusca to have arrived in the Atlantic
at the time stated. However, they must have reached
the Atlantic much later—not till pleistocene times—
if we adopt the above-stated suggestions as to the
age of the Forest-Bed (cf. p. 125). Moreover, the
great similarity between the faunas of Southern
Spain and North-western Africa indicate that the
formation of the Straits of Gibraltar is of very
recent date. The northern mollusca, of course,
could not have reached Sicily till later. To suppose
that the Sicilian deposits have been uplifted 7000 feet
since then is no doubt contrary to all our geological
teaching, but we must remember that this is altogether
an exceptional case. The area in question has probably ever since been in the immediate neighbourhood
of an active volcano, and the rate of the uplift has
therefore been immeasurably greater than at other
localities with which this one might be compared. The
disconnection between Tunis, Sicily, and Southern
Italy was evidently produced by a subsidence of the
tract of land uniting these countries. If we suppose

that this happened in early pliocene times, we have either to take for granted that the terrestrial fauna and flora of these countries are of miocene origin, or that they were joined again during the Pleistocene Epoch. The range of a very large number of animals and plants is such as can only be explained by assuming that Tunis, Sicily, Sardinia, Corsica, and Southern Italy were connected with one another. Of such extensive land-connections subsequent to the arrival of the northern marine mollusca we possess, however, no geological evidence whatsoever; and it is extremely improbable that the land-areas which had sunk were once more raised before again subsiding. The many animals whose presence in the Mediterranean Region bears witness to these ancient land-connections could not have arrived there in miocene times—in fact, they could hardly have lived there before the end of the Pliocene Epoch. On the other hand, it seems difficult to believe, once the Straits of Gibraltar were open and the waters of the Atlantic able to enter the Mediterranean, that the sunken parts between Sicily, Italy, and Tunis could have been raised without affecting the entire area of that sea. Nor is it likely that the junction between these countries could have then been brought about by a general lowering of the Mediterranean waters. As it may be asked what evidences we possess at all for the supposition of such land-connections as I have indicated, also that Southern Italy and Greece were connected, a few of the more salient instances

of distribution bearing on this problem may be of
interest.

I have already referred to the occurrence of the
remains of a small race of Red Deer in the caves
of Malta, similar to those still living in North-
west Africa, Corsica, and Sardinia. The Black-
mouthed Weasel (*Mustela boccamela*) inhabits Persia,
Asia Minor, Greece, South Italy, Sicily, and Sardinia,
while *Mustela africana* is found in Malta and Algiers.
The European Porcupine inhabits Asia Minor, the
island of Rhodos, Greece, Southern Italy, Sicily,
North Africa, and Spain. Then we have the Wild
Sheep of Asia Minor, Cyprus, Sardinia, and Corsica,
all of which are closely allied. The small shrew-like
Crocidura etrusca occurs in South France, Italy,
Sicily, and North Africa. Many other mammalia
might be quoted, but these are sufficient for our
purpose.

There are a good many reptiles and amphibians
with a similar distribution. The European Chamæleon
(*Chamæleon vulgaris*) has been found in South Spain,
North Africa, and Sicily. The Snake *Periops hippo-
crepis* is confined to Spain, Sardinia, and Greece.
The worm-like Lizard *Blanus cinereus* inhabits some
of the Greek islands, North Africa, and Spain.
Another Lizard belonging to the *Scincidæ* has also
been found in some of the Greek islands, Sicily,
Sardinia, Southern Spain, and the Canary Islands.
Discoglossus pictus—a toad—occurs in Spain, North-
west Africa, Malta, Sicily, Sardinia, and Corsica.

A variety of the Tree Frog (*Hyla arborea Savignyi*) is found in Europe only in Corsica, Sardinia, and the Greek Archipelago.

Eight species of Reptiles and Amphibia—some of which I have just referred to—are enumerated by Dr. Forsyth Major as occurring eastward and westward of the Italian peninsula (and almost all also in North Africa) without being known on the mainland of Italy. And in order to show that Sardinia and Corsica are more closely related to North Africa than to Italy, he indicates the general range of the Reptiles and Amphibians found in these islands. Of the twenty-one species, only twelve inhabit Italy, but at least sixteen North Africa and seventeen Spain. Indeed, he shows that Corsica, Sardinia, Sicily, and North-west Africa form a zoogeographical province, from which Italy, with the exception of a few localities on its west coast, is excluded. It is a remarkable fact that there are a few localities on the west coast of Italy which in their fauna and flora exhibit closer relationship with Corsica and Sardinia than with the mainland. Thus Dr. Major pointed out that the *Catena Mettalifera*, the *Monte Argentario*, and *Monte Circeo* all belong to what we may call the former Tyrrhenian continent. They are to be regarded as its eastern limits, which remained standing, while the central portion—now occupied by the Tyrrhenian Sea—subsided, and is at present covered by deep sea. Subsequently these remnants of the old continent became joined with the newly-formed Italian peninsula, but the plants and animals belonging to the older flora and fauna were mostly destroyed by newer and more vigorous immigrants. A few of the more hardy ones survived, and are a standing testimony of the geographical revolutions of that part of Southern Europe.

That the Mediterranean area has undergone such profound geographical changes as I have endeavoured to indicate is no new theory. Many zoologists who have investigated the fauna of that region, and have attempted to explain the faunistic relations, had to acknowledge that the migrations must have taken place under geographical conditions entirely different from those obtaining at present. Rütimeyer long ago remarked that it seemed to him much more probable that Morocco, Algeria, and Tunis were peopled by way of Gibraltar, and perhaps also by Sicily and Malta from Europe, than Southern Europe from Africa. After careful conchological researches in the Western Mediterranean region, Dr. Kobelt came to the conclusion that formerly Southern Spain and Morocco must have been united by a broad land-connection. Sicily and Algeria do not apparently show any very intimate relationship conchologically, but farther west—in the mountains of Tetuan—Dr. Kobelt discovered a colony of Sicilian forms.[1]

[1] There are a great many instances of discontinuous distribution among Oriental Invertebrates. Thus the Freshwater Crab (*Telphusa fluviatilis*) occurs in Southern Italy, Greece, Turkey, Cyprus, and

"The close relationship," remarks Dr. Major (*a*, p. 106), "shown in the fauna of Corsica and Sardinia to Africa, permits the supposition that the connection with these islands had persisted to a much more recent date than that with Europe."

Many other authors have pointed out the close similarity existing between the faunas of Southern Europe and North Africa. We need only refer to the writings of Professor Suess, Milne-Edwards, and Boyd Dawkins. Mr. Blanchard went even so far as to say, "a comparer les plantes et les animaux de la Sicile et de la Tunésie, on se croirait sur le même terrain" (p. 1047).

No less than 113 species of phanerogamic plants are enumerated by Professor Engler (p. 53) as occurring in the Mediterranean coast region east and west of Italy without being found in that peninsula, or at least only in the extreme south of it. But he tells us that these species represent only a portion of such plants, which are extremely numerous.

In taking a general survey of these plants, Professor Engler is of opinion that their range implies that a large number of the Mediterranean species have migrated along a line which can be drawn between North Africa, Sicily, Greece, Crete, and Asia Minor, and that from this line the distribution started northward again.

Asia Minor. Another crustacean—a Freshwater Crayfish—(*Hemicaridina Desmaresti*) inhabits Spain, Corsica, Sardinia, Sicily, and Asia Minor.

Many of these plants then, and also some of the animals I have referred to, formed part of the older stream of migration which entered Europe from Asia Minor (*vide* Fig. 5, p. 117). There were only two courses open to them as they arrived on our continent during earlier Tertiary times. They could either go straight west towards Greece, or in a more northward direction to the newly-formed Alps. As the latter were raised, some of the immigrants were modified so as to adapt themselves to the new surroundings. Others became extinct; but a great many have persisted in the Alps to the present day and exhibit discontinuous distribution, having meanwhile disappeared in the intermediate tract between the latter and their original home in Asia. The lowlands of Eastern and Central Europe were either occupied by the sea or by large freshwater lakes, so as effectually to prevent a direct migration northward.

When the newer migrants arrived from Asia not only had the Alps risen to a lofty mountain chain acting as an effectual barrier, but Southern Italy and Greece had become disconnected. Some time after, Sicily and Southern Italy also became separated. Meanwhile the stream of migrants which consisted less and less of typically southern forms, emigrants from Central Asia and even Southern Siberia, mingled with the southern forms on their way to Europe, and these now poured across the newly opened plain of Central and Northern Europe. But it was not until some time after this that the Mediterranean Sea broke across the Ægean region, and that the Northern Sea retired from the plains of Eastern Russia to admit the typical Siberian fauna and flora into our continent (*vide* pp. 189-241).

I cannot close this chapter without referring to the active distributional centre—or I might say, centre of origin—of species situated in South-eastern Europe. No group of animals is more instructive in elucidating the paths of migration from this centre than the terrestrial mollusca. Wherever the original home of the genus *Clausilia* may have been in early Tertiary times, it is certain that the most active centre of origin is now, and has been for a considerable time past, in South-eastern Europe. One of the earliest migrants from that modern centre of this interesting genus is *Clausilia bidentata*, which is the only species found in Southern Spain, and one of the two met with in Ireland, and which has been observed in high altitudes in the Alps and in Scandinavia. As we go eastward from Western Europe the number of species of *Clausilia*, as we have seen, increases until we reach a maximum in the Balkan peninsula and the region of the Caucasus. *Limax, Agriolimax,* and *Amalia,* three genera of slugs, likewise appear to have originated in the same region and spread over Europe from there. Some species like *Limax maximus* and *L. marginatus* are very ancient, and probably commenced their wanderings in early Tertiary times. In this manner many animals of European origin have joined the Oriental migrants in their westward and also in their later northward travels. In a similar way species of plants and animals of Alpine origin might have joined these migrants in their northward course, and it is only when we come to carefully analyse the constituent parts of all these members which have come to us in England from the south, that we realise the complexity of their origin. Finally, even the Siberian migrants mingled with the later Oriental ones, and in some cases the decision as to whether a certain species belongs to the former or to the latter migration becomes a matter of great difficulty.

SUMMARY OF CHAPTER VI.

LIKE the last chapter, this deals with the Asiatic migrants. But while the former described the history of the northern invasion, those animals which entered Europe from the south-east are here more particularly referred to. They originated in Central, Southern, and Western Asia. It is not easy to discriminate in all cases between this Oriental migration and the Siberian. To a certain extent, even an entry of Northern Asiatic species has taken place by the southern route, and *vice versâ*. On the other hand, southern species might have come to Europe by the southern route—that is to say, to the south of the Caspian—and also by the northern, which lay to the north of that great inland sea. The Red Deer is a good example. It arrived on our continent by both routes. However, there is a racial difference in the members of the two migrations. The small race now found in Corsica, Sardinia, North-west Africa, and Western Europe, is probably the older of the two, while the larger one—resembling the American Wapiti Deer—arrived very much later from Siberia.

The Mammoth, Wild Boar, Badger, the Dippers and Pheasants, are all Oriental species which have come to us from the south-east; but there are also Reptiles and Amphibians, and a host of Invertebrates. Not all the animals, for instance, which have reached us in England from the south-east are of Asiatic origin. There is an active centre of distribution in South-eastern Europe itself, from which species radiate out in all directions. This fact is well illustrated by the genus *Clausilia*. Species from this centre, and also from the Alps, joined the Oriental stream in their northward course.

In reviewing a number of instances of Oriental species in Europe, one is struck by the peculiarity of their having apparently followed two distinct routes. All entered from Asia Minor, which is proved to have been connected with Greece until recent geological times. From here some seem to have proceeded straight west, others northward. Further study reveals the fact that the first route was followed by a much older set of migrants at a time when the Mediterranean area was greatly different from what it is at the present day. Greece was then joined to Southern Italy, Sicily, and Tunis. The latter was also connected with Sardinia and Corsica, and the Straits of Gibraltar did not exist. Under such geographical conditions a direct migration on land from Southern Greece to Spain was not only possible, but was actually undertaken by a very large number of Oriental species.

CHAPTER VII.

THE LUSITANIAN FAUNA.

UNDER the Roman Emperor Augustus, the Spanish peninsula was divided into three provinces, one of which—namely Lusitania—occupied a large portion of the present area of Portugal. The term "Lusitanian" is therefore almost synonymous with Portuguese, but it has frequently been applied by zoologists and botanists in a much wider sense, so as to vaguely include the extreme south-west of Europe without any definite limits. Neither do I propose to restrict the term to everything found within the borders of Portugal. For the sake of convenience, we may designate as Lusitanian forms those animals and plants which have migrated to Central, Southern, or Northern Europe from South-western Europe. They may really be North-west African species, or they may have originated on land which lay to the west of Portugal, and which is now mostly buried beneath a deep sea. Nevertheless, we have received them from the extreme south-western portion of our continent —they have come to greater Europe from that direction.

In discussing the component elements of the British

fauna and flora in the third chapter, I have already
referred to the distinguishing characters of the Lusitanian migrants and to their distribution. I need
only repeat, therefore, that these are now principally
confined to the south-western portions of the British
Islands. The late Edward Forbes was the first to
trace the Lusitanian flora to its native home. In his
classical memoir on the geological relations of the
existing fauna and flora of the British Isles, he laid
the foundations of a new method of research. We are
as yet only beginning to realise the far-reaching
conclusions obtainable by a careful study of the
geographical distribution of animals and plants,
though the lines of investigation were indicated
by him more than fifty years ago. Forbes was
of opinion that the Lusitanian element in the British
flora was of miocene age, and that it survived the
Glacial period on a now sunken land to the south-
west of Ireland. Mr. Carpenter and myself agree
in so far that we are both inclined to look upon
this Lusitanian flora and the accompanying fauna
in Ireland as of pre-glacial origin. But I am
not quite satisfied that the Lusitanian migration
ceased to come north then. It may have received
a temporary check; but the presence, for instance,
of the Dartford Warbler (*Melizophilus undatus*) in
the south-east of England would seem to indicate
that its northward migration took place in very
recent times. It is possible also that the very
restricted occurrence of the Dartford Warbler may

imply that it is gradually withdrawing towards its centre of origin from a former wider range. Such an eventuality, as we have seen, has actually taken place in a great number of instances.

It is not only in the British Islands that we perceive the influence of the Lusitanian element. Scandinavia, Russia—indeed almost every part of Europe—can boast of some migrants which have originated in South-western Europe or on the mysterious lands which lay beyond it. As a rule, however, we notice a marked decrease of Lusitanian species as we travel eastward from Western Europe. Nevertheless, certain forms have travelled far beyond the confines of our continent, and we certainly meet with them in Asia and Northern Africa.

It is remarkable that we are apt to mistake sometimes for Lusitanian migrants species which are of Oriental origin. In a previous paper I classed such animals which had apparently originated in South-western Europe, but had really come from Asia by a circuitous southern route, with the Lusitanians. However, there is really no reason why the two should not be kept apart, provided we can discriminate between the pseudo-Lusitanians and the true ones. I have already indicated in the last chapter how these pseudo-Lusitanian migrants originated.

Supposing an Oriental species had left Asia for Europe in miocene times, it would on its arrival in Greece have had to decide between two courses. It could either advance into the newly-formed Alpine

19

peninsula and there remain, or at once push on westward into Southern Italy, Sicily, and Tunis, by means
of the old land-connections, and thence into Southern
Spain. The Atlantic communicated at that time with
the Mediterranean across the valley of the Guadalquivir; but that connection ceased to exist towards
the end of the Miocene Epoch, when the Oriental
migrants were free to ramble through Spain and the
whole of the North European plain. I have indicated
on a previous occasion (*a*, p. 484) that the earliest
members of the Red Deer migration, which have left
their traces in the caves of Malta, and whose descendants still live in Corsica, Sardinia, and North Africa,
may have found their way to Northern Europe in
this manner. Many other Asiatic mammals probably
reached the British Islands in a similar way.

I cannot call to mind any large species of
mammal which we might reasonably suppose to
have originated in South-western Europe. Even
among the smaller ones, few give us any definite
clue in this respect. For instance, the present range
of the genus *Myogale*—a small Insectivore belonging
to the Mole family (*Talpidæ*)—teaches us nothing.
The two living species show discontinuous distribution, and are almost confined to Europe. *Myogale*
occurs fossil in French miocene deposits, but is
unknown beyond the confines of our continent. It is
therefore probably of West European origin. The
gap between the South Russian *M. moschata* and the
Spanish *M. pyrenaica* is bridged over in so far as we

know from fossil evidence that the former had a
much wider range in pleistocene times, being then
found in England, Belgium, and Germany. *Talpa*,
too,—to which genus our common Mole belongs,—
seems to be a West European genus, since it occurs
in French miocene deposits. However, it would be
difficult to name many more recent genera which could
be included in the area which I propose to investigate
in this chapter. The genus *Lepus* is probably not of
Lusitanian origin, but the sub-genus *Oryctolagus*—to
which our common Rabbit belongs—has no doubt had
its original home in that region. Only two species of
Lepus (*Oryctolagus*) are known, one of which—*Lepus
lacostei*—has been met with in French pliocene
deposits. The other is the Rabbit (*L. cuniculus*).
Though generally considered to have been introduced into the British Islands, no reason can be
brought forward in favour of such a supposition,
especially as it is known to have spread into
Germany in pleistocene times from South-western
Europe. It occurs in France, the Spanish peninsula,
North-western Africa, and on some of the Mediterranean islands. Its nearest living relatives, as we
should almost expect, are found in South America.

Of the Lusitanian Birds I have already mentioned
the so-called Dartford Warbler (*Melizophilus undatus*),
which ranges from the south of England to the
extreme south-west of Europe. A second species
occurs on the Balearic Islands and on Corsica,
Sardinia, and Sicily. The Andalusian Bush-quail

(*Turnix sylvatica*) is probably of North African origin, and has subsequently spread into Southern Spain and Portugal, and eastward as far as Sicily. It is an instance of a migrant utilising the old Mediterranean land-connections in the opposite direction from that described in the last chapter.

Two of our British Wagtails are very closely related, so much so that it requires a very critical eye to distinguish them even at close range. They also frequently interbreed. In their distribution, however, there is a considerable difference between the White Wagtail (*Motacilla alba*) and the Pied Wagtail (*M. lugubris*). While the former ranges almost all over Europe and Asia, the latter is a local form resident in the British Islands, Southern Scandinavia, and France, and a winter visitor to Spain and North-west Africa. The genus *Motacilla* is probably Oriental in its origin, but it seems as if the Pied Wagtail was a Lusitanian species which had gradually spread northward, only to return to South-western Europe in severe weather for shelter.

The Bearded Titmouse (*Panurus biarmicus*)—the only representative of the family *Panuridæ*—may possibly be a Lusitanian bird. The fact of its being absent from Scandinavia and Northern Russia is suggestive of a southern origin. It is doubtful whether the bird occurs on the south side of the Mediterranean, but it is common in the south of France and Spain, and has also been observed in Sicily, Greece, and Asia Minor. In Central Europe it is found sparingly, and eastward its range extends as far as Turkestan.

The genus *Fringilla*, which belongs to the great family of the Finches, appears to be not only of European origin, but, if the range of the species counts for anything, I should feel inclined to locate their home in the south-west. Altogether, five species are known. One of them, viz., *Fringilla teydea*, is confined to the Island of Teneriffe; another, *F. madeirensis*, is found in Madeira, the Canaries, and the Azores; a third, *F. spodiogenys*, inhabits North-west Africa. The two remaining species have a much wider range. *F. cælebs*—the common Chaffinch—occurs in Europe, while its range extends eastward to Western Siberia, Persia, and Turkestan. The other—*F. montifringilla*, known as the Brambling—is more common in Northern Europe, and generally frequents the more northern latitudes of Asia as far as Japan.

It might be urged that the peculiar little blue Magpie of Spain—*Cyanopolius Cooki*—should find a place among the Lusitanian species, since there is no bird like it anywhere else in Europe. But in Eastern Siberia there lives a bird so closely allied as to be barely distinguishable from it. Nevertheless, since there are some distinguishing characters, it has received a distinct name—*C. cyanus*. This is a most interesting and remarkable case of discontinuous distribution, which may perhaps be explained by the supposition that the genus is of Oriental origin, and

has died out at its former headquarters in Southern
Asia and all along the line of migration, except at
the extreme limits of the range in both directions—
east and west.

As we go down in the scale of life—among the lower
vertebrates and invertebrates—we meet with a greater
number of prominent members of the Lusitanian
migration. The Bullfinch, Dipper, and Chough,
which might be thought to be of Lusitanian origin,
are, as I have shown in the last chapter, Asiatic.

The European snakes seem to be all of eastern
origin, unless *Tropidonotus viperinus* might be claimed
as a Lusitanian form. Of very great interest from
a zoogeographical point of view is our only European
member of the South American and African family
Amphisbænidæ. This species—*Blanus cinereus*—is
of the size and shape of an ordinary earth-worm,
from which, however, it may be distinguished by its
snake-like wriggling motions. It lives under stones
in Spain and Portugal, North-west Africa, and
Greece. It has, therefore, a somewhat similar distribution to that of many of the animals and
plants referred to in the last chapter. But here
we have an animal which has evidently utilised the
old Mediterranean route described on p. 271, from
west to east. Two other species of *Blanus* inhabit
Asia Minor and Syria, but most of its nearest relations
either live in South America or tropical Africa. In
migrating to North and West Africa, its ancestors
probably made use of the land-bridge which spanned

the Atlantic in early Tertiary times. An
Lusitanian Lizard—belonging not to an abc
group, but to the typical Lacertidæ—is *Psa
dromus hispanicus.* It is rather variable in col
generally of a brown or green—and grows to a l
of about four or five inches. It occurs throu
the Spanish peninsula and also in Southern F
One of the handsomest European Lizards,
reaches almost a foot in length,—of an olive c
with greenish or mother-of-pearl reflection, anc
two yellow stripes along each side of the bod
an allied species (*P. algirus*). From the S[
peninsula it passes into Southern France and
Africa. Two other species of the genus are co:
to North-west Africa.

It is quite possible that the genus *Pelobates*
south-western origin. Of the two known spec
this genus of Toads, one is found in the C
European plain and the other on the Spanish
sula and in France. The closely allied *Pe*
punctatus, too, is confined to this south-w
district, and their nearest relations are fou
Mexico. Similarly, the genus to which the M
Toad (*Alytes obstetricans*) belongs may ha
original home in that part of Europe. Of th
species, one is confined to France, Switz
Belgium, and Western Germany, and the othe
Alytes cisternasii, to Spain. *Discoglossus pi*
well-known and conspicuous Toad in Sc
Europe—inhabits Spain, Algiers, and Tun

islands of Malta, Sicily, Sardinia, and Corsica. From
the general range of the family *Discoglossidæ*, as
given in Mr. Boulenger's excellent catalogue, it
appears that nowhere in the vast space between
China and New Zealand has any member of the
family been discovered. The peculiar genus of
Salamander—*Chioglossa*—is quite confined to the
Spanish peninsula.

The Butterflies *Nemeobius lucina* and *Charaxes
jasius* may also have had their home in that south-
western district. To this migration also seems to
belong the genus *Gonepteryx*, which has so peculiar a
range in the British Islands. The only British species,
known as the Brimstone Butterfly (*Gonepteryx rhamni*),
occurs in the south of England and in the south and
west of Ireland. It is met with over the greater part
of Europe, and its range extends into Asia Minor and
Northern India, and then it reappears again in distinct varieties in Japan and the Amur district. Three
other species of *Gonepteryx* are known from Tibet
and India, and one (*G. cleopatra*) from Southern
Europe and Northern Africa. All the remaining
species inhabit the west, viz., Brazil, Mexico, and
Venezuela. That the genus has migrated from
America eastward to Europe appears to be more
probable than a migration in the opposite direction.
At any rate, that an exchange of species between
the south-western portion of the Holarctic Region
and the Neotropical area took place is indicated by
the fact, not only that a variety of *G. cleopatra* has

been found in Madeira, but also that the Canary Islands possess a distinct form of *Gonepteryx*, viz., *G. cleobule*.

Dr. Kobelt has given us such an exhaustive memoir on the characteristic Mollusca of the different zoogeographical provinces of Europe, that we are particularly well informed as regards that group of Invertebrates. He tells us that the group *Torquilla* of the genus *Pupa*—which is a small chrysalis-like snail—is especially characteristic of the Pyrenees, Spain, and Portugal. In a certain measure they replace there the *Clausiliæ* which, as we have seen in the last chapter, have come from the east and are almost entirely absent in the south-west of Europe. Of about seventy species of *Torquilla*, the larger number are confined to this district, and some, which like *Pupa (Torquilla) granum*, range eastward, have travelled along the old Mediterranean highway, *viâ* Algiers, Sicily and Greece, to Asia Minor. They are still found along the whole of this route.

Similarly, we are told by the same author, that *Gonostoma*—a group of the large genus *Helix*—has a number of species in the same south-western district, while only one, viz., *Helix obvoluta*, occurs in England and Germany, and two in the Alps. Southward we again find many representatives crossing over to North Africa, among which *Helix lenticula* has a similar range to *Pupa granum*, which I have just referred to. The Alpine sub-genus *Campylæa* is quite absent in the Lusitanian district.

Among our own British testaceous Land Mollusca, several *Helices*, viz., *Helix pisana, ericetorum, virgata, acuta, fusca, rotundata, aculeata*, and probably many others, have come to us from the south-west. The species of *Hyalinia* are undoubtedly of very remote origin, and it would be futile at the present state of our knowledge to speculate as to their home. Some of our species may possibly be of British origin. *Balea perversa* is probably a south-western species, and certainly *Pupa anglica*, which is quite confined to Western Europe.

FIG. 18.—The Spotted Slug (*Geomalacus maculosus*).

Much more characteristic of South-western Europe, however, than these land-shells are some of the slugs. The peculiar genus *Geomalacus* is almost entirely confined to Portugal. One species, which I have had several occasions to refer to in illustration of the term "discontinuous distribution," ranges far beyond the confines of that country. This is *Geomalacus maculosus* (Fig. 18), first discovered in the south-west of Ireland, and more recently also in Portugal. Although careful search has been made for it in other parts of the British Islands, this slug has only been found in the portion of Ireland just indicated. Within the last few years I have taken it, up to a height of over a thousand feet, on the promontory north of the Kenmare River, also from sea-level up to a considerable height near Glengariff, and more recently Messrs. Praeger and Welch discovered it in abundance near the town of Kenmare. But beyond this rather circumscribed area in the counties of Cork and Kerry it does not occur (*vide* Fig. 19). Several Portuguese species of this interesting genus have since been added to science by Dr. Simroth and others. Dr. Simroth, too, has promulgated the view that the genus *Arion*—to which our common brown garden slug belongs—is of Lusitanian origin. Indeed, the number of species of *Arion* diminishes as we leave that province, though one extends beyond the borders of Europe into Siberia. The same number of species, viz. five, occur in Germany and in England. *Testacella*—a slug-like mollusc—which lives underground on earthworms, and of which genus three species, viz. *T. maugei*, *T. haliotidea*, *T. scutulum*, are known to inhabit the British Islands, is another Lusitanian animal. All the species are confined to Western Europe and North Africa; they do not even reach Germany or Switzerland.

I have had occasion to mention once before an extremely interesting genus of blind Woodlouse, viz., *Platyarthrus*. Like *Testacella*, it lives underground, and also resembles it in its general range. Its distribution is therefore of particular interest. It is difficult to conceive that *Platyarthrus*, from its peculiar mode of life, could have crossed any formidable barrier, such

FIG. 19.—Map of the British Islands on which the geographical distribution of *Geomalacus maculosus* is indicated in black.

as even a narrow straits of sea. Its occurrence in Spain and North Africa indicates, therefore, that the Straits of Gibraltar did not exist at the time when it undertook the migration southward, just as the English Channel and the Irish Sea could not have been there when it wandered to England and Ireland. The species which occurs in the south of England has a wide range in Ireland, and reaches in Scotland its most northern European limit of distribution. *Platyarthrus* is only one of the Lusitanian genera of woodlice. In Ireland—chiefly on the west coast—we also find a brilliantly coloured Woodlouse, which is absent from Great Britain, viz. *Metoponorthus cingendus*. It reappears again on the Continent in the south of France. Its range is therefore suggestive of a Lusitanian origin ; and indeed, when we examine the general distribution of the genus *Metoponorthus*, we find that out of the forty-four known species, fully one-half are confined to Western Europe and North Africa.

My friend and colleague, Mr. Carpenter, informs me that among the Irish Spiders he is acquainted with, the following are to be looked upon as Lusitanian species :—

Dysdera crocota.	Agroeca celans.
Oonops pulcher.	do. gracilipes.
Tegenaria hibernica.	Teutana grossa.
Theridion aulicum.	Cnephalocotes curtus.
Lasæola inornata.	Porrhomma myops.

Of the *Coleoptera*, the genera *Trichis*, *Glycia*, and *Singilis*, all belonging to the Running Beetles (*Carabidæ*), are almost confined to the Spanish peninsula.

The beetles *Rhopalomesites Tardyi*, *Eurynebria complanata*, and *Otiorrhynchus auropunctatus* also belong to this fauna, as also the Earthworms *Allolobophora veneta* and *A. Georgii*, and the Millipede *Polydesmus gallicus*.

It will be evident to every one from these few instances of Lusitanian species, that somewhere in South-western Europe and North-western Africa, and also, perhaps, in a larger now submerged western land-area, there existed an active centre of development, from which animals spread in all directions.

If the presence of *Platyarthrus* in North-west Africa proves that the Straits of Gibraltar had come into existence after its southward migration, it also suggests that the ancestral home of this woodlouse was in the Spanish peninsula. Whether this supposition is correct or not, does not affect the Straits of Gibraltar problem, for in a migration northward into Spain from Morocco a land-connection would be equally necessary. Almost every group of vertebrates and invertebrates furnishes instances of species which must have crossed the Straits on dry land. Many naturalists have come to this conclusion, and have clearly expressed their views on the subject. At the commencement of the present period, says Mr. Bourguignat (p. 354), the north of Africa was a peninsula of Spain, the Straits of Gibraltar did not exist, and the Mediterranean communicated by the Sahara with the Atlantic.

The faunas of North-west Africa and the south-

western portion of our continent are so closely
related, that an uninterrupted intercourse by land
must have existed for a very long period. The
Mediterranean, however, throughout the Tertiary
period—at any rate since miocene times—must have
had almost constant communication with the Atlantic.
According to Professor Suess, this was the case. The
Atlantic was joined with the Mediterranean across
the valley of the Guadalquivir during the Miocene
Epoch, so that Andalusia must have belonged to
North Africa in those days. The Straits of Gibraltar
are supposed to have been formed in the next epoch.
I have already expressed my disagreement with that
theory from a zoogeographical point of view. The
old Guadalquivir connection probably persisted much
longer,—though interrupted by temporary periods of
a partial retreat—so as to uncover sufficient land to
allow of an interchange during miocene as well as
pliocene times between the European and North
African faunas. It is in this way, perhaps, that some
of the members of the Alpine fauna have reached
Spain by way of Corsica, Sardinia, and North-western
Africa, and *vice versâ*. The Balearic Islands were
then connected with Spain; and we find there many
curious survivals which have long ago become extinct
on the mainland.

That the Straits of Gibraltar are only of recent
formation has been suggested on zoogeographical
evidence by Bourguignat, Simroth, Kobelt, and many
others. Dr. Kobelt believes that the former land-

connection between the south of Spain and Morocco was much wider than is generally assumed, and that the coast-line stretched from Oran in Algeria straight across to Cartagena in Spain (*b*, ii., p. 228).

My allusions to the lands lying beyond the Lusitanian province, refer chiefly to the Canary Islands and Madeira. Whatever doubts Dr. Wallace had on the subject of their former connection with Morocco, it cannot be denied that they used to be of much larger extent, especially in miocene and pliocene times. It seems extremely probable that these islands formed part of the mainland of North Africa until comparatively recently, and that they are the last traces of a sunken continent which united Africa and South America. A discussion of this problem, however, must be deferred, as it is a complicated one, and one which would lead me altogether outside the scope of this little volume. I hope I shall have an opportunity to publish my views on this subject before long, meanwhile the reader must content himself with this mere statement.

During the greater portion of the Miocene, and I think for part of the Pliocene Epoch too, the advance of the Lusitanian species eastward was barred on the continent of Europe by an arm of the sea which stretched northward along the Rhone valley from the Mediterranean. The Lusitanian forms which originated in Southern Spain were able to travel east during these times by way of North-west Africa,

Sicily, Southern Italy, and Greece; and it is possible

FIG. 20.—The Strawberry-tree (*Arbutus unedo*) in its native habitat in the south-west of Ireland. (From a photograph by Robert Welch.)

that some may have reached the Alps in this manner, and Eastern Europe generally. That the Lusi-

20

tanian centre was never a very active one compared
with, for instance, the Oriental is indicated by many
distributional facts. It is difficult to understand,
however, why the Oriental species, on the whole,
have migrated so far west, while few Lusitanians have
gone very far east. This seems to have been noted

FIG. 21.—The Irish Spurge (*Euphorbia hiberna*) in its native habitat
in the south of Ireland. (From a photograph by Robert Welch.)

particularly in the case of the flora. Mr. Bonnet drew
attention to the fact that in Tunis there are none of the
absolutely characteristic plants of Morocco and Spain,
while the Oriental flora is represented by a good
many species. Lusitanian species have spread chiefly
southward into North Africa, and northward into
France, the British Islands, and even Scandinavia.

As I have mentioned in the third chapter, there are a good many species of Lusitanian origin in the British Islands. However, we have only a mere remnant of what we ought to have, had the climate been less trying. It is probable, too, that the submergence destroyed a good many plants and the insects dependent on them. That the Lusitanian fauna is very ancient in the British Islands is proved by the fact of the discontinuous distribution of so many species. A greater number survived in Ireland than in England.

Altogether—and this was strongly urged by Edward Forbes—the Lusitanian element is the oldest of the components of our fauna, and it must have poured into the British Islands for many geological periods almost without cessation. The same author, in his classic essay, refers especially to the Lusitanian flora, two prominent members of which are the British plants, *Arbutus unedo* (Fig. 20, p. 305) and *Euphorbia hiberna* (Fig. 21, p. 306). The former has a wide range in the Mediterranean region, and occurs in the British Islands only in the south-west of Ireland. The Spurge, on the other hand, is also found in the south-west of England, besides Ireland and Southern Europe.

SUMMARY OF CHAPTER VII.

The term "Lusitanian" is in this chapter employed in the wide sense, as indicating the South-west of Europe and North-western Africa. From this centre, and probably also from a now sunken land which lay to the west of it, issued a fauna and flora of which we have abundant evidence in our own islands, especially in Ireland. Edward Forbes held that the Lusitanian element of the British flora was of miocene age, and that it survived the Glacial period in this country.

At the time when the Straits of Gibraltar did not exist, and when there was free land communication between Asia Minor, Greece, and Tunis, many Oriental species migrated westward by this ancient Mediterranean route as far as Spain. They would then have invaded the more central parts of Europe from the south-west, without however being of Lusitanian origin. Of the true Lusitanian mammals a typical example is the Rabbit. Then we have a few birds and several interesting reptiles and amphibians. The genus to which the Brimstone Butterfly belongs is also of south-western origin. A number of Mollusca are mentioned which from their range likewise indicate a Lusitanian origin. Most of our British Slugs and many of our larger Snails belong to this group.

All these are merely a small remnant of what we received from South-western Europe during the Miocene and Pliocene Epochs. But they spread into many parts of Europe, and a few even crossed into Asia. The antiquity of the Lusitanian element in our fauna is especially indicated by the frequent recurrence of "discontinuous distribution" among the species belonging to that section.

CHAPTER VIII.

THE ALPINE FAUNA.

WE are told by Sir Archibald Geikie (p. 851) that "from the Pyrenees eastwards, through the Alps and Apennines into Greece, and the southern side of the Mediterranean basin, through the Carpathian Mountains and the Balkan into Asia Minor, and thence through Persia and the heart of Asia to the shores of China and Japan, a series of massive limestones has been traced, which, from the abundance of their characteristic foraminifera, have been called the Nummulitic Limestone. Unlike the thin, soft, modern-looking, undisturbed beds of the Anglo-Parisian area, these limestones attain a depth of sometimes several thousand feet of hard, compact, sometimes crystalline rock, passing even into marble, and they have been folded and fractured on such a colossal scale that their strata have been heaved up into lofty mountain crests sometimes 10,000, and in the Himalaya range more than 16,000 feet above the sea." "Nowhere in Europe," continues the same author (p. 860), "do oligocene strata play so important a part in the scenery of the land, or present on the whole so interesting and full a picture

309

of the state of Europe when they were deposited, as in Switzerland. Rising into massive mountains, as in the well-known Rigi and Rossberg, they attain a thickness of more than 6000 feet." "By far the larger portion of these strata is of lacustrine origin. They must have been formed in a large lake, the area of which probably underwent gradual subsidence during the period of deposition, until in Miocene times the sea once more overflowed the area."

From these remarks by our most eminent British geologist, we gather that in early Tertiary times much of the present area of Switzerland was either a sea or a large freshwater lake. The Alps were then appearing in this sea, probably as a chain of islands, and in the beginning of the Miocene Epoch one large elongated island had made its appearance—the future European Alps. I have already mentioned that the Miocene Sea skirted the Alps from the Mediterranean up the valley of the Rhone and along its northern and eastern margin. Miocene marine deposits are also known from the Southern Alps and the east side of the Apennines, from Corsica, Sardinia, and Malta. No trace, however, of them has been noticed anywhere along the Ægean Sea or on the Balkan peninsula. The Alps were therefore connected to the east with the outliers of the Balkan Mountains, and in this way with Asia, from which they received so large a proportion of their fauna and flora. In pliocene times the sea still washed the southern shore of the Alps, but to the north dry land gradually supplemented the sea, and the Alpine fauna and flora were able to pour into the plain. It was then that the Arctic species—which we have learned had migrated into Northern Europe from the north—found their way to the Alps. In a similar way Lusitanian forms—in fact, species from almost all parts of Europe—were now free to wander to the newly opened up peninsula which had become part of the mainland of Europe. The typical Siberian species had not entered our continent at that time, it was not till much later—not until the middle of the Pleistocene Epoch—that they made their appearance at the foot of the Alps, but, as we shall see later on, it is doubtful whether many of these species ever reached the mountains.

The fauna of the Alps, and also the flora, is therefore made up of a number of component elements. In the first place we have the Oriental element —the migrants from Central and Southern Asia. When the nature and origin of the Oriental fauna in Europe was discussed, reference was made to the fact (p. 272) that we can distinguish an older from a newer Oriental migration. Both of these have entered the Alps. As we might anticipate, many of the older Oriental migrants have developed into new species, laying the foundation of an indigenous Alpine element. From the fact that they set foot on the Alpine peninsula, it might be expected that there could have existed no mountains to speak of. The climate was mild and damp. Now as the country rose, and a formidable mountain range took the place of a hilly island, the whole fauna was lifted up and transferred to entirely different conditions. A modification of their structure to suit the new surroundings was therefore to be anticipated, and that is exactly what occurred, though not in all cases.

Take, for example, the goats which are of Asiatic origin. Every one has heard of the " Steinbock,"—the Alpine mountain goat (*Capra ibex*)—though very few have seen it in its native haunts, where it is now on the verge of extinction. A closely allied species (*Capra sibirica*) inhabits the Altaï and Himalayan Mountains; a third species (*Capra sinaitica*) lives in Palestine, and has entered Egypt by way of the Sinaitic peninsula. Another (*C. ægagrus*) occurs in Asia Minor, Persia, the island of Crete, and some of the Cyclades. This exemplifies what I remarked in the last chapter about the former land-connection between Greece and the Asiatic continent. Finally, we have the Pyrenean Goat (*Capra pyrenaica*), which is found in the Pyrenees, the higher ranges of Central Spain, in Andalusia, and Portugal, thus indicating that it probably reached the Spanish peninsula from the south by means of the old Sicilo-Algerian highway, especially as remains of the species occur in the cave deposits of Gibraltar. The ancestors of the goat-like Antelope—known as the Chamois (*Rupicapra tragus*)—no doubt also came from Asia. The genus is not represented there, but *Nemorhœdus* and *Budorcas* are allied Asiatic genera, while the Rocky Mountain

Goat (*Haploceros montanus*) also has certain affinities with the Chamois. Besides the Alps, the latter occurs in the Caucasus and the Pyrenees. The Alpine Marmot (*Arctomys marmotta*) is sometimes quoted as owing its origin to the Siberian pleistocene migration, but it does not occur in Siberia now, nor is there any palæontological evidence that it was ever found there. . The genus *Arctomys* is an ancient Asiatic genus, to judge from its general range. Only two species occur in Europe, one of which, the true Siberian Marmot (*A. bobac*), just enters our continent in the east—or rather, it is one of those species which came to us in pleistocene times and are now gradually retreating towards their native land. The genus, however, is probably not of Siberian origin. No less than seven other species occur in Asia, six of which are confined to Central Asia and the Himalayan Mountains, while four have wandered to North America. The sequence of events, therefore, was that the ancestor of *Arctomys marmotta* probably came to the Alps direct from Central Asia by way of Asia Minor in miocene or pliocene times. It has since become modified into a distinct species, and has spread to the European plain, where it occurs fossil in pleistocene strata, and to the Carpathian Mountains and the Pyrenees.

The great majority of species of the large genus *Microtus* (*Arvicola*) are Asiatic, and there can be little doubt that it has originated in that continent. There is one species of Vole (*Microtus nivalis*) which occurs

in the high Alps, and which has been supposed to be a
typical Alpine form. It is known, however, to occur
also in North Italy and in Bohemia, while *Microtus
leucurus* of the Pyrenees is identical with this
species. But its range is by no means confined to
Europe, for it has also been discovered in Syria
and Palestine, while a closely allied form exists in
the Himalayan Mountains. This shows clearly that
the species has migrated to the Alps from Asia
Minor. That this migration may have taken place
at an early period—at a time when Sardinia and
Corsica were still connected with Southern Europe—
is indicated by the occurrence of an extinct Vole
(*Microtus brecciensis*) in Sardinian and Corsican
pleistocene (?) deposits.

All the Alpine species mentioned except the
Chamois can be easily traced to their former Asiatic
home. But even it has its nearest relations in Asia.
I might also refer to another Vole (*Evotomys Nageri*)
which is practically confined to the Alps and
Northern Italy, and which has probably originated
there, though most of its nearest relations are either
Asiatic or North American species.

But besides these Asiatic immigrants and their
modified descendants we have a small truly native
Alpine mammalian fauna. *Sorex alpinus*—the Alpine
Shrew—occurs only in the Alps, the Harz Mountains,
Pyrenees, and Carpathians. The genus has been
found in European eocene strata,—in vastly older
deposits in our own continent than elsewhere,—so

that it is extremely probable that it has originated there. It may then have developed a new centre of distribution in the newly-formed Alps, where both *Sorex alpinus* and *S. minutus (pygmæus)* have their home. From there they again spread—perhaps already in miocene times—to Asia and North America, where a large number of new species originated. It seems to me even probable that one of these Asiatic species of *Sorex*, viz. *S. araneus (vulgaris)*, subsequently migrated towards the old home of its forefathers, since we find it more or less confined to Central and Northern Asia and Northern Europe.

Though the origin of the Alpine Hare has already been referred to and fully discussed in a previous chapter (p. 148), the conclusions arrived at may be once more repeated. The Alpine Hare *(Lepus variabilis)* is of Arctic origin. It spread southward into Europe, North America, and Asia in early glacial times, and reached our continent from Spitsbergen by means of a direct land-connection with Lapland. The Scandinavian peninsula was then separated from Russia, but connected with Scotland and Ireland (Fig. 13, p. 170). Since England was then united to France, the Alpine Hare was able to invade western continental Europe and all the mountain ranges. Its range is very discontinuous, small colonies being scattered all over the mountainous parts of the Northern Hemisphere, while the European Hare—a closely allied species—occurs in the plain, and now occupies to some extent the former haunts of the Alpine Hare (cf. Fig. 8, p. 137). Might not the European Hare, as suggested, possess some advantages which enabled it to drive the other into more inaccessible parts, thus producing the peculiarity of range? The present distribution of the Alpine and the European Hare (*L. Europæus*) appears to me to strongly support such an assumption. It is not the cold which has driven the Alpine Hare to the Alps; and its presence there is not, as is often supposed, a "*standing testimony of a former arctic climate*" in Europe, but merely the necessary consequence of the weaker species being thrust into less accessible regions by a stronger rival.

Muscardinus avellanarius,—the common Dormouse, —though by no means confined to the Alps, has probably originated there. It is found up to a height of nearly 5000 feet in these mountains, and is spread over Europe at nearly equal distances from the Alps in all directions. Being absent from Ireland, Scotland, Norway, and Northern Russia, it seems as if it had only diffused northward in more recent times.

The closely allied genus *Myoxus* is likewise of European extraction, some species being known from French eocene deposits.

There are only a few typically Alpine Birds. One of these is the Alpine Accentor (*Accentor collaris*), which on rare occasions visits England, and Northern Europe generally. It is, however, by no means peculiar to the European Alps; a variety of this species occurs in Central Asia, Eastern Siberia, and Japan. The only other Accentor inhabiting our continent is the Hedge Accentor (*A. modularis*), which is resident over the greater part of it, and also in North Africa and the Mediterranean Islands. It also extends its range across the Ægean Sea to Asia Minor, so that really not a single Accentor is peculiar to Europe.

Both the European species are evidently old forms, and the genus, as might be expected, is certainly Asiatic. No less than ten other species of Accentor are known, all of which are confined to Central Asia and the Himalayan Mountains, and are therefore all Holarctic. I may mention that much difference of opinion still exists as to the true zoological position of this anomalous genus. It has been located in several different families by various ornithologists, but has not yet found a permanent resting-place. Another bird generally considered to be peculiar to Switzerland is the Alpine Chough (*Pyrrhocorax alpinus*), but its range extends across Asia Minor to the Himalayas. Whether the European Chough should not form a distinct genus is a matter of opinion. Some of our leading ornithologists, like Dr. B. Sharpe, are inclined to separate it from *Pyrrhocorax;* however, there is no doubt that it is closely related to the Alpine Chough, whatever view we may take of the generic distinctness. It inhabits principally Western and Southern Europe, also

North Africa; and its range extends eastward to
the Himalayas, China, and Eastern Siberia. If any
doubt still existed as to the Asiatic origin of the
Choughs, it may be noted that the only two other
closely allied genera, viz., *Corcorax* and *Podoces*, live
in Australia and Central Asia respectively.

There are two other birds to which I should like
to refer. These are the Rock Sparrow and the Alpine
Snow Finch. The first of these (*Petronia stulta*) is
by no means peculiar to the Alps. It is the only
species of the genus inhabiting Europe; and besides
the Alps it occurs in Southern Europe generally,
and ranges as far west as the Canaries and Madeira.
Eastward it is not found beyond Central Asia. Of
the remaining five species of *Petronia*, two occur in
Asia (including India) and three in Africa. Whether
the genus is African or Asiatic is immaterial for our
purpose, since, in any case, the only European species
came to us from the east with the Oriental migration.
The distribution of the Alpine Snow Finch (*Montifringilla nivalis*) is very similar to that of the birds we
have just been considering. It inhabits the Alps up
to a great height, but occurs also on the Pyrenees and
other South European mountain ranges as far east as
Palestine, where again it is found in the Lebanon.
The genus *Montifringilla* has seventeen other species.
Twelve of these live in Central Asia and Japan,
extending as far north as Kamtchatka, while five
inhabit Western North America right down to
Mexico. There is every probability that in this case

also we have to deal with an Asiatic genus which spread eastward to America, and westward to Europe. As regards the Reptiles, there are *no* peculiar Alpine forms, but among the Amphibia some species deserve to be mentioned. Up to an elevation of 10,000 feet we find in the Alps the Black Salamander (*Salamandra atra*); and it is apparently quite peculiar to them, never having been observed in the plains. The handsome black and yellow Salamander (*Salamandra maculosa*)—so well known as a terrarium specimen—likewise occurs in the Alps, and it has besides a fairly wide distribution in Europe. It is known from Southern Germany, the Pyrenees, Spain, Portugal, Sardinia, Corsica, Greece, Syria, and Algiers. A third species (*S. caucasica*) inhabits the Caucasus. The evidence of distribution here points emphatically to an Alpine origin of the genus *Salamandra*. We cannot tell where the ancestors of *Salamandra* may have come from, but several other genera of *Salamandridæ* are certainly Asiatic. Our common Newt (*Molge vulgaris*) belongs to a genus with nineteen species, several of which are peculiar to Europe. The general range of the genus, however, extends to North America, and it is more probable therefore that it originated in Asia. If so, it certainly must have passed into Europe at a very early date. Let us assume the first *Molges* to have traversed the Ægean Sea on *terra firma* to Greece in miocene times, they might thus have been able to travel straight on to the old Tyrrhenian continent of which Corsica and

Sardinia now form the remains, and also on to North-west Africa. Indeed, we find high up in the Corsican mountains an interesting large brownish-grey Newt (*Molge montana*), and another in Sardinia (*Molge Rusconii*). Again, in Algiers there are two species, viz., *Molge Poireti* and *M. Hagenmülleri*, while the Moroccan *M. Waltlii* passes into the south of Spain. Here *Molge boscæ*, *M. aspera*, and *M. marmorata* originated, the latter passing into France.

Another branch of the *Molge* tribe turned north-ward from Greece towards the newly forming Alps; and there originated *Molge alpestris* and *M. palmata*, which more recently have spread into England (one at least), Germany, France, Austria, and Southern Italy. *Molge vulgaris* is an Asiatic species which wandered northward after entering Europe, covering a large area, but never reached the extreme south or south-west. *M. cristata*—the large Water Newt—has a similar but not quite so extended a range, while *M. vittata* never managed to cross the borders of Asia Minor. Some of the other species occur in China, Japan, and North America.

None of the tailless Batrachians—the Frogs and Toads—are peculiar to the Alps, but one, viz. *Rana temporaria*, ascends to the height of no less than 10,000 feet. It is our common British Frog. No other Frog probably ranges so far north or to such heights.

Let us now inquire what the invertebrate fauna of the Alps teaches us. We are told by Dr. Kobelt,

the great authority on European land shells, that a
uniformity of character marks the Alpine Molluscan
fâuna (*b*, i., p. 251). One of the characteristic genera
Campylaea—often looked upon as a sub-genus of
Helix—is a group containing somewhat flattened
conspicuous snails of large size. These are found
everywhere in the Alps, and wherever they occur
beyond the confines of these mountains, remarks Dr.
Kobelt, their origin from the main stock is easily
traced. They have been gathered in the Apennines in Sicily, and even beyond the Mediterranean in
Algeria. On the Balkan peninsula they occur right
down to the most southern point of Greece, but
are not met with either in Crete or Asia Minor.
One species has been found sub-fossil in Thuringia
in Northern Germany.

Another truly Alpine genus, says Dr. Kobelt, is
the operculate *Pomatias*, which in its geographical
distribution offers some interesting modifications from
that of *Campylaea*. Less limited to high elevations,
it has spread over a greater part of the plains. This
has happened especially in France, while in Germany
one species advances almost as far north as Heidel-
berg. In other directions also the genus has travelled
beyond the limits of range of *Campylaea*. *Pomatias*
occurs in the Pyrenees and Northern Spain, in
Sardinia and Crete, and may, according to the same
author, be expected in Asia Minor, although no
species has as yet been met with there. In Greece,
again, it has been observed, and numerous species

inhabit Tunis and Algeria. Dr. Kobelt connects the
wider range of *Pomatias* with the geological history
of the genus (*b*, i., p. 253). He tells us that species of
Pomatias have been found in eocene deposits differ-
ing but little from our present forms, while undoubted
Campylaeae are not met with till we reach the upper
Miocene.

Zonites is, according to Dr. Kobelt, a third Alpine
genus, whose range scarcely differs from the other
two (*b*, i., p. 254). The centre of distribution lies at
present in one of the branches of the most southern
Alpine chain which help to form a large portion of
the Balkan peninsula. The bulk of the species
inhabit that peninsula, the Greek Islands (except
Crete) and Asia Minor. Neither in the Tyrol nor
in Switzerland do we find any *Zonites*, and the few
species that do occur in the south-eastern Alps
only just cross the outliers of these mountains.
Between the south-western Alps and the Rhone we
again find a *Zonites*—a remarkable case of discontinuous distribution, since the nearest other habitat
of the genus is Monte Gargano in South-eastern
Italy, which is known to harbour a good many
interesting geographical puzzles.

We still have a good deal to learn as regards the
molluscan fauna of Sicily, Sardinia, and Corsica.
These islands have scarcely been more than skimmed
by conchologists, and *Zonites* may inhabit one or all
of these, which might indicate to us the manner in
which this genus travelled from Southern Italy to

Provence in the south of France. The distribution of *Zonites* certainly does not seem to imply an Alpine origin, because it is almost completely absent from the Alps proper. But I do not think my views differ materially from those of Dr. Kobelt, since the Alps, in the wide sense, include the mountains of the Balkan peninsula, where I should feel inclined to locate the ancestral home of the genus.

The small operculate genus *Acme* is a similar case. Dr. Kobelt places the centre of distribution on the southern slope of the Alps, but scarcely any of the species inhabit the Alps proper. Some occur in France, others in North Africa, Sicily, Southern Italy, and the Caucasus. It is evidently a very ancient genus. The species live in moss or underground, and are not likely to be transported across the sea by accidental or occasional means of distribution.

Still another genus, which resembles *Acme* in its geographical distribution, is *Daudebardia*—a small slug-like mollusc with a tiny shell. It does not, however, range nearly so far north or west as *Acme*, for it occurs neither in the British Islands nor in Spain or the Pyrenees.

I shall not be able to refer to more than a few of the most typical Alpine species of Lepidoptera, but they may be taken as fair examples of the geographical distribution of the rest of the group.

Even those visitors to Switzerland who do not claim to be naturalists have heard of the remarkably

handsome and stately Butterfly known as Apollo.
To the ardent entomologist, the first sight of this
typical Alpine species is a never-to-be-forgotten delight, and he generally brings home with him a rich
harvest of specimens. The more experienced Butterfly hunter knows that there are no less than three
different kinds of Apollo—or, as we should say more
correctly, of Parnassius—in Switzerland. There is
first the common Apollo (*Parnassius Apollo*), then
the rarer and more local *P. delius*, which inhabits
more elevated regions, and finally the still scarcer *P.
mnemosyne*, which is only known from the highest
mountain ranges. It may be a surprise to those who
have accustomed themselves to connect Apollo with
the Alps, and who think the two belong together and
cannot do without one another, to hear that it is
by no means confined to them. It is also found in
Scandinavia, France, Spain, Russia, and in Siberia.
Parnassius delius is confined to the European Alps
and the mountains of Central Asia, while *P. mnemosyne* is known from the Pyrenees, Sweden, Hungary,
Sicily, Russia, and Western Asia. One other Par-
nassius inhabits Europe, viz., *P. Nordmanni* of the
Caucasus, but all the remaining species of the genus
—and there are nearly thirty more—are confined to
Central Asia. A few, as we have seen, have reached
Europe, some have travelled to the Himalayan
Mountains, and others to Western North America.
The centre of distribution is certainly in Central
Asia, and we have no reason to suppose that the

original home in this case does not agree with that
çentre.

Melitæa, a genus to which some of our British
Fritillaries belong, has also some typically Alpine
members. Two of these, viz. *M. cynthia* and *M.
asteria*, are peculiar to the Alps, the latter being
only found at considerable elevations. Most of the
remaining fourteen European species are also found
in Central Asia. Thus the isolated *M. maturna*,
which in Europe is confined to Lapland, is also
known from the Altaï Mountains, which again are
near the centre of distribution, since some species
of *Melitæa* range across the Northern Pacific to
Western North America.

The small British Mountain Ringlet, and also the
Scotch Argus, belong to a genus of butterflies which
is very characteristic of the European Alps. But
owing to its enormous geographical distribution, its
probable home is somewhat difficult to ascertain.
Nevertheless it is a noteworthy genus, especially so
from the fact that the two British species *Erebia
epiphron* and *E. æthiops* are taken at first sight
for true Arctic migrants. As neither of them, however, occurs in Scandinavia, Greenland, or Arctic
America, this supposition must be abandoned. They
must be looked upon as species which once had a
wider range in the southern parts of the British
Islands, and which have survived in a few isolated
localities, where they are apparently on the verge of
extinction.

About sixty species of *Erebia* are known to science, half of which are found in Europe, the remainder in Siberia, the Himalayas, Arctic America, Chili, Patagonia, South Africa, and Madagascar. Though a few do range into these outlying regions of the earth, Central Asia seems to lie near the centre of distribution of the genus, and the probability is that it also was its original home. Most of the European species are high Alpine forms—*E. glacialis* being met with at a height of 10,000 feet—and these are generally quite peculiar to the Alps, showing that their ancestors came from Asia at an early date, probably by way of Asia Minor and Greece. A few, as for instance *E. lappona*, range right across to the Altaï Mountains from the Alps, and at least one—*E. melas*—is found in Greece. *Erebia* migrations seem therefore to have taken place by the Southern or Oriental route at different geological periods. But some of the European species which are more or less confined to the plain, and are either absent from Switzerland or do not reach the higher elevations, appear to me to have come by the more direct northern or Siberian highway, at a still more recent period. These are *Erebia æthiops, medusa, ligea,* and *ambla*.

Only one species of the well-known Polar genus *Œneis*, viz. *Œ. aëllo*, occurs in the Alps. It has always been taken at very high elevations near the verge of the snow-line on the most lofty parts of the Simplon Pass, and other similar situations.

Altogether about a dozen species of this genus of butterfly are known, most of which are confined to the polar regions of the Old World and the New, though some have found their way to the extreme south end of South America, in what manner is still a mystery. Like the preceding genera, this also appears to have emerged from Central Asia. The genus, too, is closely allied to the last, and though its range is not quite so extensive, it resembles it in many respects. The Alpine species of *Œneis* came to Europe by the Oriental route. But the Lapland species—at any rate *Œ. jutta* and *Œ. bore*—have taken a somewhat circuitous route to reach our continent. They first migrated from Asia to North America, and then by the old land-connections by way of Greenland to Lapland. It is noteworthy that Professor Engler felt convinced (cf. p. 171) that the occurrence of many of the Arctic plants in North Scandinavia and Siberia could be best explained by the assumption of such a migration from Asia *viâ* North America to Europe rather than by the shorter route.

There are far more Alpine beetles than butterflies, but their geographical distribution is less well known, and it is therefore not at all safe to base important conclusions as to the origin of a fauna on that group alone; however, as far as my limited knowledge of the *Coleoptera* of the Alps goes, their general range seems to agree perfectly with other orders of insects. Many can also be traced to an Asiatic home, and

the route they came by is the Oriental and not what
I have called the Siberian.

Take, for instance, the genus *Nebria*, of which we
have one species in England—a black insect with a
bright reddish-yellow border and long light legs—
known as *N. livida*. There are about eighty European species, most of which are confined to the Alps,
the Caucasus, the Pyrenees, Spain, and Greece. The
genus, however, ranges all over the Holarctic Region,
that is to say roughly, over Europe, Central and
Northern Asia, and North America. The centre of
distribution lies in Central Asia. If the genus had
poured into Europe by the northern or Siberian
route, we should probably now find many species in
Northern Russia, Germany, and France; but this is
not the case, and we may therefore assume with some
justification that the Southern or Oriental route was
the only one available at the time when the bulk of
the species of *Nebria* wandered to Europe. Many
of the *Nebrias* occur in Switzerland and in the Alps,
generally on the margins of the snow-fields and
glaciers, like *N. Germari* and *Brunii*. Others, for
example, *N. atrata*, ascend to the highest limit of
animal life, having been observed at a height of over
10,000 feet.

Of the remaining orders of insects we know as yet
very little. Central Asia and even Siberia are only
beginning to be explored, and their invertebrate
fauna—except *Lepidoptera* and *Coleoptera*—is practically unknown. However, I cannot conclude this

short summary of some of the more characteristic
Alpine animals without referring to the Grasshoppers
which are so conspicuous in the mountains. The
mountain air simply rings during a bright summer's
day with the loud and cheerful song of millions
of these insects. It is one of the most vivid impressions a tourist brings back from Switzerland—this
constant shrill sound issuing from an apparently
invisible source.

Among these Grasshoppers there are some highly
characteristic Alpine genera. *Pezotettix*—formerly
known as *Podisma*—is one of these. *P. alpinus* is
almost confined to the high Alps; with *P. mendax*
it occurs in lower levels chiefly towards the south-
east, that is to say, in the direction of Hungary,
Servia, and Dalmatia. *P. frigidus* occurs not only
in the high Alps, but also in Lapland. *P. Schmidti*
and *P. salamandra* are found in Carinthia, Servia,
and Transylvania; and one species also inhabits
the Pyrenees and another the Italian Mountains.
Finally, the only English species of *Pezotettix*,
viz. *P. pedestris*, has been taken in Sweden, Denmark, and then again in Austria, Hungary, Servia,
etc., as far east as the Volga, and also on the high
Alps, in Sardinia and the Abruzzi Mountains in
Italy.

Very little, as I remarked, is known of the Asiatic
range of this genus, but either the same or a closely
allied one has many representatives in North and
South America. Whether *Pezotettix* is therefore

of Asiatic origin we cannot positively affirm, but whatever view we take, the general range of the European species indicates that the migration took place from the Alps in a south-easterly direction, or to them in a north-westerly one. That is to say the Oriental route, and not the Siberian, was utilised by the migrants.

Fortunately, we know a little more about another Grasshopper genus, called *Chrysochraon*. There are only two species, one of which, *Chr. dispar*, has been found from Northern France to the mountains of Servia, but not in the Alps. The other, *Chr. brachypterus*, has a somewhat similar range in the plain; but, moreover, it inhabits the Alps up to a considerable height. It is interesting to note that both these Grasshoppers again turn up on the Amur in Eastern Siberia.

In conclusion, I might mention one more Grasshopper, viz. *Tettix*, because it includes a species—*T. bipunctatus*—which, though well known in the plain of Middle and North Europe, ascends the Alps to a height of nearly 10,000 feet. It is one of the few instances I know of an animal occurring in the same form in such an enormous range of altitude —from sea-level to the highest regions where animal life is known to exist. It is also known from Asia Minor and Siberia. *T. subulatus* has a similar distribution, but is more common in Southern Europe than the other. *T. fuliginosus* occurs in Lapland and Siberia, *T. meridionalis* and *T. depressus* all

along the shores of the Mediterranean. There can be
no doubt that here also we can trace migration to or
from Siberia, and again, as on previous occasions, by
the Oriental route.

We now possess a fair general idea of the fauna of
the Alps. We have learned that a good many of
the animals are indigenous, and that others have
migrated to the Alps by various routes. The majority
of these have come from Central and Southern Asia
with what has been described as the Oriental migration. A much smaller number have reached the
Alps from the north and the west, but none of the
latter are among the high Alpine forms. What will
be the most surprising revelation is that the eastern
species, which arrived in Europe with the Siberian
migration, are practically absent from the Alps
proper. No doubt some of them still survive in the
lowlands of Switzerland and the Tyrol, but none of
the true Alpine fauna owes its origin to the Siberian
migration. If we compare the Alpine mammals with
the Siberian forms which reached England (*vide*
p. 202), we at once perceive the difference. We
should expect to find in the Alps—if not the Reindeer and the Glutton—the Arctic Fox, the little
Pica, the Lemmings, and the pouched Marmots. It
might be urged that some of the smaller Siberian
carnivores and rodents do inhabit the Alps. So they
do. The Stoat and Weasel have found such a con-
genial home in Europe, both in the plain and
mountains, that they have spread rapidly to the

latter, and no doubt reached within a comparatively short time the great heights at which they now occur in the Alps. But the Voles (*Arvicola*) have scarcely spread beyond the region of fields and cultivated ground. A height of 5000 feet at the most marks their maximum altitude in the Alps.

The fauna which reached the Alps in miocene and pliocene times, as well as the indigenous element, must have survived the Glacial period in their mountain home. Though I think that the conditions of the climate at that time and the size of the Scandinavian glaciers have been greatly exaggerated, there can be no doubt at all about the enormous size of many of the Alpine glaciers at this period. The climate was probably much moister but not colder than what it is now, possibly warmer. The snowfall was therefore greater, so that glaciers filled many of the lower valleys of Switzerland which are now quite free from ice, and even invaded the plain. But there is no reason whatsoever why the Alps should not even then have supported a luxuriant fauna and flora as they do now. Possibly many of the miocene plants and animals became extinct then, but extinction of species occurs at the present day. We hear complaints that the Chamois and the Steinbock have nearly vanished; we know that the Marmot is now much scarcer than it used to be, and that the Edelweiss and many other plants are more and more difficult to find, and seem rapidly to disappear. No doubt all this is in a great measure due to the influence of man, but not altogether. There is a constant struggle for existence going on among the animals and plants themselves—the stronger and fitter species driving the less fit and weaker into a corner, where they finally succumb. This happens now just as it did in pliocene and pleistocene times, and need not imply change of climate.

As soon as the Miocene sea to the north of the mountains had retreated, a portion of the Alpine fauna poured into the plain, and many species have found their way to the British Islands, a few to Scandinavia and Russia. Westward too, the sea soon after retired and opened a way for those Alpine species which were vigorous enough to extend their range in that direction. South-eastward, of course, a highway had long ago been open, and Alpine forms which were able to migrate towards the incoming Oriental stream, had no difficulty in doing so. When they arrived in Greece, some turned westward again and populated Sicily, Southern Italy, Sardinia, Corsica, and Northern Africa, while others crossed over to Asia Minor, which was then connected with Greece, and wandered towards the Central Asiatic or the Himalayan Mountains.

But, as I remarked, few of the typical Alpine species reached Scandinavia and Lapland. I have already referred to the similarity between the Northern Scandinavian and the Alpine faunas in a previous chapter, and I have shown that this resemblance

cannot altogether be explained by the supposition of
an interchange in the faunas of the two regions. That
this has taken place to some extent is probable, but
the resemblance appears more especially due to the
fact that the Alps and Scandinavia have been peopled
from the same centres of distribution.

In order to make this matter quite clear, I will give
a familiar example as an instance of the manner in
which the present distribution can be explained without taking recourse to direct migration from the Alps
to Scandinavia or *vice versâ*. The example I will
take is that of a family of birds, not only of extreme
interest from the fact of its northern range, but also
from the pleasure it gives to those addicted to sport.
This is the grouse family, the *Tetraonidæ*.

Let us commence with our British Grouse (*Lagopus
scoticus*), which is peculiar to the British Islands.
In Norway we find a Grouse (*L. albus*) which differs
in habit, and in the fact of its turning white in
winter; otherwise it is so closely allied to our
Grouse that many ornithologists do not separate
them specifically. No doubt the British Grouse is a
descendant of this Scandinavian Willow-grouse. The
latter is known also to inhabit Greenland and Arctic
North America, and it is even found beyond Behring
Straits in Northern Siberia. *En route* between
Scandinavia and Asia, travelling in a westward
direction, we meet with two other very local species
of Grouse, which may be looked upon as offshoots
of *L. rupestris*—viz., *L. hyperboreus* of Spitsbergen,

and *leucurus* of Western North America. In Asia we
then again find two kinds of Grouse, very closely
related, and by some indeed regarded as belonging
to the same species. These are *L. rupestris* and
L. mutus. Mr. Ogilvie-Grant tells us of the former
(p. 49), that it is merely a more northern rufous form
of *L. mutus*, and that it goes through similar changes
of plumage. In summer the males are readily distinguishable, but in winter it is impossible to tell
one from the other. "*L. rupestris* taken as a whole,"
says Mr. Ogilvie-Grant, "appears to us barely specifically distinct from *L. mutus.*" *L. rupestris* occurs
not only in Northern Asia, but crosses the Behring
Straits to Arctic America, being still found on the
Aleutian Islands, which represent the last remains
of the former land-bridge between Asia and North
America, then eastward to Greenland and Iceland.
However, while this form does not cross the confines
of Asia in a westerly direction, its near relative
L. mutus—better known as the Ptarmigan—does;
and may perhaps have entered Europe as a Siberian
and also as an Arctic migrant. It is still found in
the Ural Mountains, in Finland, and the highlands
of Scandinavia. It is gradually being driven out of
the Alpine lowlands, while it has long ago disappeared from Germany, France, and Austria—in
fact, from all the lowlands of Europe. It has also
been met with in the Pyrenees and in some of the
Spanish mountains. Similarly, the bird has become
extinct in England and Ireland, while it is becoming

more and more scarce in Scotland. The centre of distribution of the genus lies in Arctic America, and from there the genus has spread to Europe and Asia. *L. albus* and *L. mutus* appear in our continent chiefly as Arctic migrants.

The Black Grouse (*Lyrurus tetrix*) belongs to a closely allied genus, which has only two species. One of these is very local in distribution, being confined to the Caucasus, but the smallness of range is to some extent compensated for by the peculiarity of its name, which is *L. mlokosiewiczi*. The Black Grouse, on the contrary, is widely distributed. It inhabits Northern Asia from the Pacific to the Ural Mountains, and extends as far south as Pekin and the Tian Shan range. In Europe it is found from the extreme east to the Pyrenees, the Apennines on the south, and to Great Britain and Scandinavia in the north. It is important to note its absence from Spain, the Mediterranean islands, and Ireland; and we have learned that it is one of those Siberian migrants which have succeeded in establishing themselves in the Alps.

The Capercaillie (*Tetrao urogallus*)—another great favourite with sportsmen—is now generally separated generically from the Black Grouse, though they are of course near relations. Its range greatly resembles that of the Black Grouse, except that it does not go quite as far east in Siberia, not having been met with beyond Lake Baikal. From there it is found westward as far as the Pyrenees. It occurs also in the Carpathians and the Alps. In England, where it used to be known by the name Cock of the Wood, it became extinct at some remote period in history, while it lingered on in Scotland and Ireland until the end of the last century. In Scotland it has been reintroduced into several counties, and being protected, it appears to spread from these artificial centres of distribution.

Like the Black Cock, the Capercaillie is a Siberian migrant, and it is one of the few Siberian species which have reached Ireland, as I have had occasion to mention in dealing with the origin of the British fauna. Two other species of Capercaillie and an allied genus (*Falcipennis*) are met with in the extreme north-east of Siberia, and six other genera, all belonging to the grouse family, are confined to North America. We have therefore a very intimate relationship between the grouse of Asia and those of North America, some species even ranging right across the two continents.

The last genus of this very interesting family is *Tetrastes*. This grouse is not familiar to British ornithologists, since it is entirely absent from the British Islands. But sportsmen who have tramped over Scandinavia know it well by the name of Hazel Grouse. It is ashy grey in colour, barred and vermiculated with black. The Common Hazel Grouse (*Tetrastes bonasia*) is found from Northern Spain in the west right through the mountainous parts of Central and Northern Europe and Northern

22

Asia to Kamtchatka and the Russian convict island
of Saghalien in the Pacific. Besides the Common
Hazel Grouse, two other species are known, one from
Eastern Russia and the other from China.

Having now shortly reviewed the whole grouse
family, we have seen that, although some species
live within the Polar Circle, the majority are
more or less confined to the more temperate
or rather the less arctic parts of the Northern
Hemisphere. They are quite absent from Southern
Asia and even the southern parts of North America,
and almost so from the Mediterranean basin. The
whole range of the family is therefore suggestive of
a northern origin, and this view agrees perfectly with
all the details of distribution. The centre of distribution lies in Northern Asia, or in Arctic North
America. From there the great genus *Lagopus*
spread east and west, reaching Europe by these
vastly divergent routes at a time when the physical
geography was very different from what it is to-
day. Several of the species common to the Alps
and Scandinavia have migrated from Siberia direct
to Eastern Europe. But we can now imagine how
from a similar centre in Asia—perhaps at a rather
more remote time—a species spread eastward across
North America and Greenland to Scandinavia, and
westward along the mountain ranges of the Tian
Shan and the mountains of Asia Minor to Greece,
and finally to the Alps. We should then have the
same species in the Alps and in Scandinavia, not far

removed from one another; but how different were
their paths of migration! This, however, is not an
imaginary instance. Such a migration must have
actually taken place in a good number of instances
among the terrestrial invertebrates and also among
plants.

The view still current among many zoologists and
botanists, that animals and plants were driven down
into the plain from the mountains of Europe during
the height of the Glacial period and there lived
together till the return of a more genial temperature,
when they retreated to their mountain homes, is a
very plausible one. During their sojourn in the
plain, the plants and animals—say from Scandinavia—intermingled with those from the Alps; and
when the time of separation came, many Alpine
forms retired northward with the Scandinavians,
while many Scandinavians would go with the
Alpines to their home. In this way the similarity
between the Alpine and Scandinavian faunas and
floras is assumed to have been brought about.
These theories, first promulgated by Edward Forbes,
were hailed with general satisfaction by the scientific
world. Even Darwin says of them (p. 331), that
grounded as they are on the perfectly well-
ascertained occurrence of a former Glacial period,
they seemed to him to explain in a satisfactory
manner the present distribution of the Alpine and
Arctic productions of Europe. To the present day
this view meets with much favour among naturalists.

It is somewhat similar to one which has recently been strongly supported by Professor Nehring and accepted by Professor Th. Studer and many others. They have never made it quite clear whether the pre-glacial fauna and flora are supposed to have been absolutely destroyed by the glacial climate, or whether part of them have been able to take refuge somewhere in the south; but the great mass of our Alpine plants and animals are believed to have been derived from the Siberian invasion, which I have fully described in the fifth chapter. This invasion spread over the European plain, and when the climate ameliorated, both animals and plants migrated north and south to the mountains. This view agrees with the earlier theory, except that the adaptation to Alpine conditions would, according to the former, have taken place since the close of the Glacial period, during which time no such modification or change of species seems to have been produced in other parts of the world. The characteristic fauna of the Alps, as has been gathered from the preceding pages, is mainly of Central Asiatic rather than of Siberian origin. Migration to the Alps took place by the Oriental route long before the Siberian invasion. Some of the species of the latter have penetrated to the Alps, but these Siberian species have not given to the fauna of the highest European mountain range the striking character with which we all associate it.

Before concluding this chapter, a few remarks on

the botanical aspect of the Alpine problem might not be out of place. It will enable us to judge which of the views indicated is the more probable, and will add' to the interest which may have been aroused by the perusal of this sketch of the fauna of the Alps. Very much the same train of argument was applied as to the course of events in the formation of the Alpine flora as in the case of the fauna. The plants were all supposed to have been killed or driven away by the arctic temperature of the Glacial period, and their place taken by new migrants from the north or east when the climate ameliorated.

Professor Engler, one of the highest living authorities on the geographical distribution of plants, is of opinion (p. 102) that a large number of the indigenous Alpine species did not originate till after the close of the Glacial period, because so many of them are absent from the Sierra Nevada in Spain, where the condition for their well-being exists, while many have evidently spread from the Alps to the Carpathian Mountains and to the Pyrenees. He does not believe that a glacial flora could have existed in the plain between the Sierra Nevada and the Pyrenees during the Glacial period (p. 109). In speaking of the Caucasus, Professor Engler informs us (p. 117) that a good many species which do not occur in the Alps reached these mountains from Siberia. Apart from the northern glacial plants, the Caucasus has only few species in common with the Alps, more with the Balkan moun-

tains and Northern Persia. Turning to Afghanistan,
our author mentions (p. 121) a few Alpine plants as
occurring in that country, and likewise in the
Caucasus and the Himalayas. He considers it
probable that the route of migration of some glacial
plants from the east to the west, and *vice versâ*, lay
across the Afghan mountains. Many of our Alpine
plants occur in the Siberian mountains, but in the Altai
and Eastern Siberia generally a considerable number
of these are by no means confined to the mountains
(p. 125). They are also met with in the lower regions,
and the rare Alpine Edelweiss (*Leontopodium alpinum*)
frequently covers wide tracts in the plain, and is
passed by almost unnoticed by the Siberian botanist.

Special attention is drawn by Professor Engler to
the fact (p. 130) that several of the Siberian plants
inhabit the Alps and the Caucasus, but are not
found in Scandinavia. And from this he deduces the
conclusion that part of the Siberian flora migrated
in a south-westerly direction towards the Caucasus
and the mountains of the Mediterranean area, exactly
in the manner indicated in respect to the fauna of the
Alps. We learned that the migration to the Alps
from Central and perhaps also parts of Northern
Asia took a south-westerly course first, and was
then followed by one in an easterly direction. I
called the former the Oriental migration and the
latter the Siberian. Later on Professor Engler
states (p. 142) that the main mass of the Siberian
forms of plants certainly wandered westward to

the south of the Ural. This is proved by the
numerous glacial plants found in the Caucasus, while
the glacial flora of the Ural Mountains is poor.
Finally, he expresses the opinion that the probability
of most of the Alpine plants occurring in Arctic
Siberia, having wandered from the Alps, by way
of Scandinavia, Greenland, and North America, to
North-eastern Siberia, is greater than the direct
migration from Europe to Siberia (p. 143).

Another continental writer on the Alpine flora who
deserves special mention is Dr. Christ. His observation that Alpine plants by no means suffer from a
high temperature (p. 309), but solely from a drying
up of the soil, seems to me to point to the correctness
of the view I have expressed on several occasions,
that these plants have originated long before the
Glacial period at a time when the climate was
warmer and moister than it is now. It seems quite
natural to Dr. Christ that the Arcto-Alpine flora
should have originated in Asia, but he excepts thirty
species which are absent from Northern Asia, though
occurring in America (p. 327). These he thinks have
penetrated direct from America to the Alps by way
of Scandinavia, since no less than twenty-three still
occur in the latter country. In the human population
of the Alps, he continues (p. 336), one can distinguish
an indigenous Celtic race, a Germanic colder and
more apathetic race, and a more lively Roman one.
The flora is composed of quite a similar mixture.
We find also an indigenous element—an Arctic and

a Mediterranean one. The last element is a survival
of the Tertiary flora of the Central European plateau
(p. 532). The plants were driven down to the shores
of the Mediterranean, and it is only after the retreat
of the glaciers that a few of them have been able to
regain their ancient territory. The incoming Asiatic
and North American flora likewise retired at the end
of the Glacial period to the Alps and the Arctic
countries, and left isolated traces of its former
abundance on the North European plain. The
bulk of the Arctic or Alpine flora is held to be
of Asiatic origin. Since Siberia shows little trace
of having been glaciated, owing to the dryness of the
climate, a rich flora was able to develop there, which
spread into Europe as soon as the vanishing glaciers
made room for it.

These are the views of Professor Engler and Dr.
Christ. They agree in so far as both of them maintain that the bulk of the Alpine flora is post-glacial
—that is to say, that it has developed quite
recently, or migrated to the Alps after the glaciers
had retreated from the plain to the mountain
recesses. It is assumed by Dr. Christ that while
Europe was practically uninhabitable, a rich flora
survived in Northern Asia, because the climate there
was too dry for the development of glaciers. Due
consideration in this interesting speculation, however,
is not given to the fact which he himself emphasised,
that Alpine plants are particularly prone to suffer
from a dry climate. Even a moderately dry cold

kills most of them. How can we then reconcile this
fact with the theory of their origin in a dry and
intensely cold climate? I quite agree with the views
as to the Asiatic origin of the bulk of the Alpine flora,
while the dry state of the Siberian climate is certainly
indicated by the extremely feeble development of the
glaciers during a large part of the Glacial period.
We know, however, that in Pliocene and even in
early Glacial times the atmospheric conditions must
have been very different in Siberia. A great slice of
Central Asia was under water, and numerous freshwater lakes covered the lowlands in the north, so
that the climate must have been damp though not
cold enough for the formation of extensive glaciers.
Everything, in fact, seems to indicate that the migration of the Asiatic Alpine flora took place at a very
early date—probably long before the Glacial period—
either by the Oriental or by the Arctic route *viâ*
North America, Greenland, and Scandinavia. But
would this not necessitate a survival of the Alpine
plants in the Alps themselves? That is the view which
has already been expressed with regard to the fauna,
and the flora probably followed a very similar course.
This is by no means a novel theory, however, and
though unfortunately an untimely death has removed
one of our very best authorities on the Alpine flora
before he had completed his life's work, we have
some indications in the earlier writings of John Ball
that his opinions on the origin of that flora did not
coincide with those held by the leading continental

authors. To quote the words of this distinguished botanist (p. 576): "Is it credible that in the short interval since the close of the Glacial period hundreds of very distinct species and several genera have been developed in the Alps, and—what is no less hard to conceive—that several of these non-Arctic species and genera should still more recently have been distributed at wide intervals throughout a discontinuous chain some 1500 miles in length, from the Pyrenees to the Eastern Carpathians? Nor would the difficulties cease there. You would have left unexplained the fact that many of the non-Arctic types which are present in the Alps are represented in the mountains of distant regions, not by the same, but by allied species, which must have descended from a common ancestor; that one species of *Wulfenia*, for example, inhabits one small corner of the Alps, that another is found in Northern Syria, while a third allied species has its home in the Himalaya." Mr. Ball is of opinion (p. 584) that the effects of the Glacial period have been greatly overrated. "Even during the period of maximum cold the highest ridges of the Alps were not completely covered with snow and ice; for we still see by the appearance of the surface, the limit above which the ancient ice did not reach, and in the middle zone the slopes that rose above the ancient glaciers had a summer climate not very different from that which now prevails. In my opinion the effect of the Glacial period on the growth of plants in the Alps was to lower the vertical height of the zones of vegetation by from one to two thousand feet." He acknowledges that there was probably a moderate diminution of the mean temperature of Europe with an increased snowfall, so as to cause a great extension of glaciers on all the mountains of Northern Europe. "But that the climate of Middle Europe was such that the plants of the high Alps could spread across the plains seems to me an improbable supposition" (p. 584).

On the Continent, also, some botanists seem to feel that Forbes's theories of the origin of the Alpine flora, which were at first hailed with such delight, and accepted by almost every naturalist as the final verdict, must be modified in the light of recent researches. Professor Krasan believes that many plants which now live in the high Alps flourished in pliocene times at sea-level (p. 37). "Especially the evergreen species exhibit the impression of an originally mild climate—of a climate without winter frosts—for otherwise the plants would have developed into species with deciduous leaves." To the favourable conditions, consisting in periodic snowfalls and high summer temperature, must be attributed the fact that in the highlands so many more species from Tertiary times have survived than in the plains. The temperature was probably much higher during the Glacial period than is generally believed. The climate was more moist, thus contributing to an abundant snowfall, while the survivors of ancient Tertiary times were able to repeople the parts which were temporarily devastated by the advancing glaciers.

In so short a chapter it is impossible to deal with the Alpine fauna in a manner more deserving of this theme. I have merely sought to give a sketch of the general outlines of the subject and to suggest another possible mode of origin of Alpine animals than that currently believed in by naturalists. It is to be hoped these suggestions will be useful to those intending to reinvestigate the problems raised in this chapter. When our knowledge of the fauna of Asia is more complete, it will be possible to give a more thorough and in many respects a more satisfactory history of the European fauna than at present.

SUMMARY OF CHAPTER VIII.

In early Tertiary times the area now covered by the European Alps was covered by the sea. Islands slowly rose above the surface of the waters, which finally coalesced to form a peninsula connected with the mainland in the east. Animals now began to invade the new territory which continued to rise, while the sea retired farther and farther to the north and south. During the Pliocene Epoch the sea ceased to wash the northern shores of the Alps, and both emigration and immigration became possible in that direction, and also from and to the west.

The Alpine fauna and also the flora are made up of a number of elements, the eastern one being the oldest. The latter is represented in the Alps by the older and newer Oriental migration. The general range of the Alpine Steinbock, Chamois, Marmot, Vole, Shrew, and Hare are specially referred to. The

Alpine birds are few in number, and all of them are readily traceable to an Asiatic ancestry. Among the Amphibia, the Salamanders are considered of Alpine origin.

Dr. Kobelt tells us that a uniformity of character marks the Alpine molluscan fauna. *Campylaea*,—often considered a sub-genus of *Helix*,—*Pomatias*, *Zonites*, are looked upon as truly Alpine genera. For very long periods the Alps seem to have received no addition to their molluscan fauna from other areas. The case is very different with the *Lepidoptera*, some of the most striking species being evidently Asiatic immigrants. Some examples of *Coleoptera* and *Orthoptera* are mentioned, and their origin discussed.

We find as the result of these considerations that the majority of the Alpine species are either indigenous or have come from Asia with the Oriental migration. None of the northern or western immigrants appear to be among the characteristic Alpine species, and it seems that the Siberian migrants have not retired to the Alps, as some naturalists have been led to suppose. It is evident that the fauna must have survived the Glacial period on the Alps, though according to geological evidence glaciers of enormous size originated on these mountains.

The identity of many Alpine species with Scandinavian ones appears at first sight due to a direct migration from the Alps to Scandinavia or *vice versâ*. Perhaps such a migration has taken place to some extent, but it is probable that from a Central Asiatic centre some species spread across Arctic America into Northern Europe, and also westward to the Alps. The Grouse family forms an interesting example.

There are two older theories which explain the similarity between the Scandinavian and Alpine faunas. Forbes's view, which gained most adherents among naturalists, was that the Scandinavian and Alpine animals were driven into the plain by the cold during the Glacial period, and when they ultimately regained their homes, some individuals of the northern species moved southward, and a few of the southern ones northward. By the more recent theory of Nehring, the Siberian animals which invaded our continent from the east, and then spread northward to Scandinavia and southward to the Alps, formed the nucleus of the faunas of these two areas. The objections to both of these views are fully set forth in this chapter.

A few remarks on the botanical aspect of the Alpine problem conclude the chapter. The origin of the flora has been explained in a very similar manner to that of the fauna. But already Ball and Krasan have raised their voices against the current theories, as the facts of distribution appear to them more satisfactorily explained on lines more consonant with those which I have used in discussing the origin of the Alpine fauna. One of the most important conclusions obtained by this study of the flora in conjunction with the fauna, is that I have emphasised in most of the preceding chapters—viz., that the Glacial period in Europe was not a time of extreme cold, and that its destructive effect on the animals and plants was by no means such as is currently believed.

BIBLIOGRAPHY.

(Titles of Works and Papers referred to in the Text.)

———•◆•———

Adams, A. Leith.—Report on the History of Irish Fossil Mammals, "Proc. Royal Irish Acad." (2nd series), vol. iii., 1878.

Alston, E. R.—"The Fauna of Scotland," 1880.

Ball, J.—On the Origin of the Flora of the European Alps, "Proc. Royal Geograph. Soc.," vol. i., 1879.

Barrett-Hamilton, G. E. H. (*vide* also Thomas and Barrett-Hamilton).—Notes on the Introduction of the Brown Hare in Ireland, "Irish Naturalist," vol. vii., 1898.

Beddard, F. E.—"A Text-book of Zoogeography," 1895.

Bell (*vide* Kendall and Bell).

Blanchard, E.—Preuves de la Formation Récente de la Méditerranée, "Comptes Rendus," vol. xciii., 1881.

Blanford, W. T.—Address delivered at the Anniversary Meeting of the Geological Society of London, 1890.

Blytt, A.—"Essay on the Immigration of the Norwegian Flora," 1876.

Boettger, O.—a. System. Verzeich. d. lebenden Arten d. Landschneckengattung Clausilia, "Ber. Offenbach Ver. f. Naturkunde," 1878. b. Entwicklung d. Pupa Arten d. Mittelrheingebiets, "Jahrbuch d. nass. Ver. f. Naturw.," vol. xlii.

Bogdanov, M. M.—Quelques mots sur l'histoire de la faune de la Russie d'Europe, "Archives des Sciences physiques et naturelles" (N.S.), vol. lvi., 1876.

Bonney, T. G.—"Ice-work, Present and Past," 1896.

Boulenger, G. A.—British Museum Catalogue (Amphibia), 1882.

Bourguignat, J. R.—Distribution géographique des Mollusques terrestres et fluviatiles en Algérie, "Annales des Sciences naturelles" (5th ser.), vol. v., 1864.

Brandt, J. F.—Zoogeographische und Palæontologische Beiträge, "Verhandl. der Kais. russ. Mineral. Gesell." (2nd ser.), vol. ii., 1867.

Brauer, A.—Die Arktische Subregion, "Zool. Jahrbücher" (Abth. System.), 1888.

Brehm, A. E.—"From North Pole to Equator," 1896.

Brooke, Sir V.—On the Classification of the Cervidæ, "Proc. Zool. Soc. London," 1878.

Bulman, G. W.—The Effect of the Glacial Period on the Fauna and Flora of the British Islands, "Natural Science," vol. iii., 1893.

Bunge, D.—La faune éteinte des bouches de la Léna et des îles de la Nouvelle Sibérie, "Congrès Intern. de Zoologie," vol. ii., Moscow, 1892.

Carpenter, G. H.—Problems of the

British Fauna, "Natural Science," vol. ii., 1897.

Christ, H.—"Das Pflanzenleben der Schweiz," 1879.

Cole, G. J.—Borderland, "Irish Naturalist," vol. vi., 1897.

Credner, R.—Die Reliktenseen, "Petermann's Geog. Mitth.," 1887.

Croll, J.—"Climate and Time," 1875.

Darwin, C.—"Origin of Species," 6th ed., 1878.

Dawkins, W. Boyd.—a. "Early Man in Britain," 1880. b. "The British Pleistocene Mammalia," part A, Preliminary Treatise, 1878. c. The Former Range of the Reindeer in Europe, "Popular Science Review," vol. vii., 1868.

Drude, O.—Betrachtungen ü. d. hypothetischen vegetationslosen Einöden im temperirten Klima der Nördlichen Hemisphäre zur Eiszeit, "Petermann's Geog. Mitth.," vol. xxxv., 1889.

Emery, C.—On the Origin of European and North American Ants, "Nature," vol. lii., 1895.

Engler, A.—"Versuch einer Entwicklungsgeschichte d. extratrop. Florengebiete d. Nördlichen Hemisphäre," 1879

Falsan, A.—"La Période Glaciaire," 1889.

Feilden, H. W.—a. Notes on the Glacial Geology of Arctic Europe and its Islands, part 2, "Quart. Journ. Geol. Soc.," vol. lii., 1896. b. Notes on a small Collection of Spitsbergen Plants, "Trans. Norfolk and Norwich Nat. Hist. Soc.," vol. vi., 1894.

Fischer, P.—Faune malacologique de Cauterets, suivie d'une étude sur la répartition des Mollusques, "Journal de Conchyliologie" (3rd series), vol. xvi., 1876.

Forbes, E.—The Geological Relations of the Fauna and Flora of the British Isles, etc., "Memoirs Geol. Survey," vol. i., 1846.

Geikie, Sir A.—"Text-book of Geology," 1882.

Geikie, J.—a. "The Great Ice Age," 1894. b. On Changes of Climate during the Glacial Period, "Geological Magazine," vol. ix., 1872.

Haacke, W.—Der Nordpol als Schöpfungs-zentrum der Landfauna, "Biologisches Centralblatt," vol. vi., 1887.

Harlé, E.—Sur la succession de diverses faunes à la fin du quaternaire, dans le sud-ouest de la France, "Societé d'Histoire Naturelle," Toulouse, 1893.

Harmer, F. W.—On the Pliocene Deposits of Holland, and their relation to the English and Belgian Crags, "Quarterly Journal Geological Soc.," vol. lii., 1896.

Herdman, W. A., and Lomas, J.—On the Floor Deposits of the Irish Sea, "Proc. Liverpool Geological Society," vol. viii., 1898.

Hofman, E.—Die Isoporien der Europäischen Tagfalter, "Inaugural Diss. d. Phil. Fakultät in Jena," 1873.

Hooker, J. D.—Outlines of the Distribution of Arctic Plants, "Trans. Linnean Soc.," vol. xxiii., 1862.

Howorth, Sir H.—"The Mammoth and the Flood," 1887.

Ihering, H. von.—a. Die Ameisen von Rio Grande d. Sul, "Berliner Entomolog. Zeitschrift," vol. iii., 1894. b. Najaden von São Paulo und die geographische Verbreitung d. Süsswasserfaunen von Süd Amerika, "Archiv. f. Naturgesch.," 1893.

Jordan, H.—Die Binnenmollusken der nördl. gemäss. Länder, "Nova Acta Acad. Carol.-Leop.," vol. xlv., 1883.

Judd, J. W.—Address to the Geological Section, "British Association Report," Aberdeen, 1885.

Karpinski, A.—Übersicht d. physiko-geograph. Verhältnisse d. Europ. Russlands, etc., "Beiträge z. Kenntniss d. Russ. Reichs" (3 Folge), vol. iv., 1888.

Kendall, P. F., and Bell, A.—On the Pliocene Beds of St. Erth, "Quart. Jour. Geol. Soc.," vol. xlii., 1886.

Kennard, A. S., and Woodward, B. B.—The Mollusca of the English Cave-Deposits, "Proc. Malacological Soc.," vol. ii., 1897.

Kew, H. W.—"The Dispersal of Shells," 1893.

Kinahan, G. H.—On possible land-connections in recent geological times between Ireland and Great Britain, "Trans. Manchester Geol. Soc.," vol. xxiv., 1896.

Kobelt, W.—a. Das Verhältniss d. Europ. Landmoll. z. Westindien und Central Amerika, "Nachrichtsblatt d. deutschen Malak. Gesell.," 1886. b. "Studien zur Zoogeographie," vols. i., ii., 1897-98.

Köppen, F. T.—Das Fehlen d. Eichhörnchens, etc., in der Krim, "Beiträge zur Kenntniss d. Russ. Reichs" (2 Folge), vol. vi., 1883.

Krasan, F.—Zur Abstammungsgesch. d. autochthonen Pflanzenarten, "Mitth. d. Naturw. Ver. Steiermark," 1896.

Lamplugh, G. W.—The Glacial Period and the Irish Fauna, "Nature," vol. lvii., 1898.

Lomas (vide Herdman and Lomas).

Lydekker, R.—a. "A Handbook of the British Mammalia," 1895. b. "A Geographical History of Mammals," 1896. c. "The Deer of all Lands," 1898. d. Catalogue of the Fossil Mammalia of the British Museum, pt. 4, 1886.

Major, C. J. Forsyth.—a. Die Tyrrhenis, "Kosmos," 7 Jahrg.,

1883. b. Studien zur Geschichte der Wildschweine, "Zool. Anzeiger," vol. vi., 1883.

Mallet, R.—Some Remarks upon the Movements of Post-tertiary and other Discontinuous Masses, "Journal Geol. Soc. Dublin," vol. v., 1850-53.

Merriam, C. H.—The Geographic Distribution of Life in North America, "Proc. Biological Soc. Washington," vol. vii., 1892.

Murray, Andrew.—"The Geographical Distribution of Mammals," 1866.

Nathorst, A. G.—Kritische Bemerkungen über die Geschichte d. Vegetation Grönlands, "Botanische Jahrbücher," vol. xiv., 1891.

Nehring, A.—"Tundren und Steppen der Jetzt und Vorzeit," 1890.

Neumayr, M.—"Erdgeschichte," vol. ii., 1887.

Ogilvie, W. R. Grant.—"Game Birds," 1895.

Penck, A.—Die grosse Eiszeit, "Himmel und Erde."

Petersen, W.—Die Lepidopterenfauna d. arktisch. Gebiete von Europa, "Beiträge z. Kenntniss d. Russ. Reichs," vol. iv., 1888.

Pettersen, K.—Arktis, "Arch. f. Math. og Naturvid.," 1882.

Pohlig, H.—Dentition und Kraniologie des Elephas antiquus (part 2), "Nova Acta Acad. Carol.-Leop.," vol. liii., 1889.

Reade, Mellard—a. The Present Aspects of Glacial Geology, "Geological Magazine," vol. iii., 1896. b. High Level Shelly Sands and Gravels, "Natural Science," vol. iii., 1893.

Reid, C.—The Pliocene Deposits of Britain, "Mem. Geol. Survey of the United Kingdom," 1890.

Rütimeyer, L.—"Die Herkunft unserer Thierwelt," 1867.

Sars, G. O.—Crustacea Caspica, "Bull. Acad. Imp. d. Sciences

St. Petersburg," vol. xiii., 1893-94.

Scharff, R. F.—a. On the Origin of the European Fauna, " Proc. Royal Irish Acad." (3rd. ser.), vol. iv., 1897. b. Some Remarks on the Distribution of British Land and Freshwater Mollusca, "Conchologist," vol. ii., 1893. c. Étude sur les mammifères de la Région Holarctique, " Mém. de la Soc. Zool. de France," vol. viii., 1894.

Schulz, A.—"Grundzüge einer Entwicklungsgeschichte d. Pflanzenwelt Mittel-europas," 1894.

Sharpe, R. Bowdler.—British Museum Catalogue (Birds).

Simroth, H.—Eine Bearbeitung d. Russischen Nacktschnecken fauna, "Ann. Musée Zool St. Petersburg," 1896.

Sjögren, H.—Ueber das diluviale Aralo-Kaspische Meer und die nord-europäische Vereisung, "Jahrb. d. Kais. K. Geol. Reichsanstalt," vol. xl., 1890.

Sollas, W. J.—On the Origin of Freshwater Faunas, " Trans. Royal Dublin Soc." (2nd ser.), vol. iii., 1884.

Speyer, A. and A.—"Die Geographische Verbreitung der Schmetterlinge Deutschlands und der Schweiz," vol. i., 1858.

Struckmann, C.—Ueber die Verbreitung d. Rennthiers, in d. Gegenwart, etc., "Zeitschr. d. deutsch. Geol. Gesell.," vol. iii., 1883.

Suess, E.—"Das Antlitz der Erde," vol. i., 1892.

Tcherski, J. D.—Das Janaland und die Neusibirischen Inseln, "Mém. Acad. Imp. St. Petersburg," vol. xl., 1892.

Thomas, O., and Barrett-Hamilton, G. E. H.—The Irish Stoat distinct from the British species, "Zoologist," 1895.

Wallace, A. R.—"Island Life," 2nd edit., 1892.

Warming, E.—Ueber Grönland's Vegetation, "Botanische Jahrbücher," vol. x., 1888-89.

White, F. B.—Some Thoughts on the Distribution of the British Butterflies, "The Entomologist," vol. xiv., 1881.

Woodward, B. B.—On the Pleistocene (non-marine) Mollusca of the London District, "Proc. Geologists' Association," vol. ii. (vide also Kennard and Woodward).

Wright, J.—Boulder-clay a marine deposit, "Trans. Geol. Soc. Glasgow," vol. x., 1896.

INDEX.

～～＊

THE END.

THE WALTER SCOTT PRESS, NEWCASTLE-ON-TYNE.

"The most attractive Birthday Book ever published."

Crown Quarto, in specially designed Cover, Cloth, Price 6s.

"*Wedding Present*" *Edition, in Silver Cloth, 7s. 6d., in Box. Also in Limp Morocco, in Box.*

An Entirely New Edition. Revised Throughout.

With Twelve Full-Page Portraits of Celebrated Musicians.

DEDICATED TO PADEREWSKI.

The Music of the Poets:

A MUSICIANS' BIRTHDAY BOOK.

COMPILED BY ELEONORE D'ESTERRE-KEELING.

This is an entirely new edition of this popular work. The size has been altered, the page having been made a little longer and narrower (9 × 6½ inches), thus allowing space for a larger number of autographs. The setting-up of the pages has also been improved, and a large number of names of composers, instrumentalists and singers, has been added to those which appeared in the previous edition. A special feature of the book consists in the reproduction in fac-simile of autographs, and autographic music, of living composers; among the many new autographs which have been added to the present edition being those of MM. Paderewski (to whom the book is dedicated), Mascagni, Eugen d'Albert, Sarasate, Hamish McCunn, and C. Hubert Parry. Merely as a volume of poetry about music, this book makes a charming anthology, the selections of verse extending from a period anterior to Chaucer to the present day.

Among the additional writers represented in the new edition are Alfred Austin, Arthur Christopher Benson, John Davidson, Norman Gale, Richard Le Gallienne, Nora Hopper, Jean Ingelow, George Meredith, Alice Meynell, Coventry Patmore, Mary Robinson, Francis Thompson, Dr. Todhunter, Katharine Tynan, William Watson, and W. B. Yeats. The new edition is illustrated with portraits of Handel, Beethoven, Bach, Gluck, Chopin, Wagner, Liszt, Rubinstein, and others.' The compiler has taken the greatest pains to make the new edition of the work as complete as possible; and a new binding has been specially designed by an eminent artist.

LONDON: WALTER SCOTT, LTD., PATERNOSTER SQUARE.

Crown 8vo, about 350 pp. each, Cloth Cover, 2/6 per Vol.;
Half-Polished Morocco, Gilt Top, 5s.

Count Tolstoy's Works.

The following Volumes are already issued—

A RUSSIAN PROPRIETOR.

THE COSSACKS.

IVAN ILYITCH, AND OTHER
STORIES.

MY RELIGION.

LIFE.

MY CONFESSION.

CHILDHOOD, BOYHOOD,
YOUTH.

THE PHYSIOLOGY OF WAR.

ANNA KARÉNINA. 3/6.

WHAT TO DO?

WAR AND PEACE. (4 vols.)

THE LONG EXILE, ETC.

SEVASTOPOL.

THE KREUTZER SONATA, AND
FAMILY HAPPINESS.

THE KINGDOM OF GOD IS
WITHIN YOU.

WORK WHILE YE HAVE THE
LIGHT.

THE GOSPEL IN BRIEF.

Uniform with the above—
IMPRESSIONS OF RUSSIA. By Dr. GEORG BRANDES.

Post 4to, Cloth, Price 1s.
PATRIOTISM AND CHRISTIANITY.
To which is appended a Reply to Criticisms of the Work.
By COUNT TOLSTOY.

1/- Booklets by Count Tolstoy.

Bound in White Grained Boards, with Gilt Lettering.

WHERE LOVE IS, THERE GOD
IS ALSO.

THE TWO PILGRIMS.

WHAT MEN LIVE BY.

THE GODSON.

IF YOU NEGLECT THE FIRE,
YOU DON'T PUT IT OUT.

WHAT SHALL IT PROFIT A MAN?

2/- Booklets by Count Tolstoy.

NEW EDITIONS, REVISED.
Small 12mo, Cloth, with Embossed Design on Cover, each containing
Two Stories by Count Tolstoy, and Two Drawings by
H. R. Millar. In Box, Price 2s. each.

Volume I. contains—
WHERE LOVE IS, THERE GOD
IS ALSO.
THE GODSON.

Volume II. contains—
WHAT MEN LIVE BY.
WHAT SHALL IT PROFIT A
MAN?

Volume III. contains—
THE TWO PILGRIMS.
IF YOU NEGLECT THE FIRE,
YOU DON'T PUT IT OUT.

Volume IV. contains—
MASTER AND MAN.

Volume V. contains—
TOLSTOY'S PARABLES.

London: WALTER SCOTT, LIMITED, Paternoster Square.

The Emerald Library.

Crown 8vo, Gilt Top, Half Bound in Dark Green Ribbed
Cloth, with Light Green Cloth Sides, 2s. each.

Barnaby Rudge
Old Curiosity Shop
Pickwick Papers
Nicholas Nickleby
Oliver Twist
Martin Chuzzlewit
Sketches by Boz
Olive
The Ogilvies
Ivanhoe
Kenilworth
Jacob Faithful
Peter Simple
Paul Clifford
Eugene Aram
Ernest Maltravers
Alice; or, The Mysteries
Rienzi
Pelham
The Last Days of Pompeii
The Scottish Chiefs
Wilson's Tales
The Fair God
Miss Beresford's Mystery
A Mountain Daisy
Hazel; or, Perilpoint Lighthouse
Vicar of Wakefield
Prince of the House of David
Wide, Wide World
Village Tales
Ben-Hur
Uncle Tom's Cabin
Robinson Crusoe
The White Slave
Charles O'Malley
Midshipman Easy
Bride of Lammermoor
Heart of Midlothian
Last of the Barons
Old Mortality
Tom Cringle's Log
Cruise of the Midge
Colleen Bawn
Valentine Vox
Night and Morning
Bunyan
Foxe's Book of Martyrs
Mansfield Park
Last of the Mohicans
Poor Jack
The Lamplighter
Jane Eyre
Pillar of Fire
Throne of David
Dombey and Son
Vanity Fair
Infelice

Beulah
Harry Lorrequer
Essays of Elia
Sheridan's Plays
Waverley
Quentin Durward
Talisman
From Jest to Earnest
Knight of 19th Century
Caudle's Lectures
Jack Hinton
Bret Harte
Ingoldsby Legends
Handy Andy
Lewis Arundel
Guy Mannering
Rob Roy
Fortunes of Nigel
Man in the Iron Mask
Great Composers
Louise de la Valliere
Great Painters
Rory O'More
Arabian Nights
Swiss Family Robinson
Andersen's Fairy Tales
Three Musketeers
Twenty Years After
Vicomte de Bragelonne
Monte Cristo—Dantes
  „ Revenge of Dantes
The Newcomes
Life of Robert Moffat
Life of Gladstone
Cranford
North and South
Life of Gen. Gordon
Lincoln and Garfield
Great Modern Women
Henry Esmond
Alton Locke
Life of Livingstone
Life of Grace Darling
White's Selborne
Tales of the Covenanters
Barriers Burned Away
Opening a Chestnut Burr
Pendennis
David Copperfield
Luck of Barry Lyndon
St. Elmo
Son of Porthos
Stanley and Africa
Life of Wesley
Life of Spurgeon
For Lust of Gold
Wooing of Webster
At the Mercy of Tiberius
Countess of Rudolstadt

Consuelo
Two Years before the Mast
Fair Maid of Perth
Peveril of the Peak
Shirley
Queechy
Naomi; or, the Last Days of Jerusalem
Little Women and Good Wives
Hypatia
Villette
Ruth
Agatha's Husband
Head of the Family
Old Helmet
Bleak House
Cecil Dreeme
Melbourne House
Wuthering Heights
The Days of Bruce
The Vale of Cedars
Hunchback of Notre Dame
Vashti
The Caxtons
Harold, Last of the Saxon Kings
Toilers of the Sea
What Can She Do?
New Border Tales
Frank Fairlegh
Zanoni
Macaria
Inez
Conduct and Duty
Windsor Castle
Hard Times
Tower of London
John Halifax, Gentleman
Westward Ho!
Lavengro
It is Never Too Late to Mend
Two Years Ago
In His Steps
Crucifixion of Phillip Strong
His Brother's Keeper
Robert Hardy's Seven Days, and Malcom Kirk (in 1 vol.)
Richard Bruce
The Twentieth Door
House of the Seven Gables
Elsie Venner
The Romany Rye
Little Dorrit
The Scarlet Letter
Mary Barton

London: WALTER SCOTT, LIMITED, Paternoster Square.

www.ingramcontent.com/pod-product-compliance
Lightning Source LLC
Chambersburg PA
CBHW030908270326
41929CB00008B/618